AUSTIN - HEALE _ 100

SERIES BNI

Service Manual

SEPTEMBER 1956

THE AUSTIN MOTOR COMPANY LIMITED, BIRMINGHAM, ENGLAND

Publication Part No. 97H997D

Introduction

THIS service manual gives comprehensive and authentic information for the assistance of the Austin Distributors and Dealer organisations in maintaining the Austin Healey 100 in perfect mechanical condition.

Full illustration of both major and sub-assemblies accompany the detailed instructions for dismantling, assembling and inspection of component parts. It is emphasised that only genuine Austin Spare parts are to be used as replacements. In addition, the service manual should not be referred to when ordering new components; the operator should at all times make use of his "Spare Parts List".

Modification which may occur on later models will be published in subsequent editions of the manual.

Although produced primarily for the use of the Austin Distributors and Dealer organisations this publication can be purchased by the car owner from his local distributor or direct from the Advertising Department, The Austin Motor Company Limited, Longbridge, Birmingham, England.

CONTENTS

100 Series BN1

100 Series BN2 Supplement

AUSTIN HEALEY '100' WITH WINDSCREEN IN RACING POSITION

AUSTIN-HEALEY '100' WITH HOOD RAISED

GENERAL INDEX

GENERAL INDEX

GENERAL INDEX

GENERAL SPECIFICATION

IN all correspondence, relating to the car, which is addressed to the Austin Motor Company Limited, or their Distributors, or Dealers, the owner should quote the engine and chassis numbers. The engine number will be found on the right hand side of the cylinder block whilst the chassis number is located on the R.H. side of the chassis frame at the front end (i.e. in the engine compartment).

ENGINE

Number of cylinders	4
Size of bore	3.4375 in. (87.3 mm.)
Length of stroke	4.375 in. (111.1 mm.)
Capacity	162.2 cu. in. (2,660 c.c.)
Brake horse power	90 (91.248 cv.) at 4,000 r.p.m.
Max. torque	150 lb./ft. (20.74 kgm.) at 2,000 r.p.m.
Compression ratio	7.5 to 1
Firing order	1, 3, 4, 2
Valves	Overhead, push rod operation
Valves springs	Double
Valve timing inlet	Opens 5° B.T.D.C. Closes 45° A.B.D.C.
Valve timing exhaust	Opens 40° B.B.D.C. Closes 10° A.T.D.C.
Valve clearance cold	.012 in. (0.03 mm.) between rocker and valve stem

LUBRICATION

Type of system	Wet sump
Oil pump	Straight gear
Oil pressure (normal running)	Not less than 50/55 lb./sq. in. (3.515 - 3.867 kg./cm.²)
Oil filter	Full flow
Filter capacity	1¼ Imp. pints (1.5 U.S. pints; 0.71 litre)
Sump capacity	11¾ Imp. pints (14.11 U.S. pints; 6.68 litres)
Pressure relief valve	Spring loaded plunger

FUEL SYSTEM

Carburetters	Twin S.U. at 20° angle
Needle model	Q.W.
Pump	S.U. Electric
Model	"L" High Pressure
Tank capacity	12 Imp. gals (14.4 U.S. gals.; 54.6 litres)
Air cleaner	Oil wetted type

COOLING SYSTEM

Radiator	Flat tube type, pressurised to 7 lb./sq. in. (.492 kg./cm.²)
Circulation	Pump assisted thermo-syphon
Fan	4 blade, belt drive
Control	Bellows type thermostat
Running temperature	185°–194°F. (75°–80°C.)
Capacity	20 Imp. pints (24 U.S. pints; 10.8 litres)

IGNITION

Coil	Lucas B.21
Distributor	Lucas D.M.2
Contact breaker gap	.014–.016 in. (.3556–.4064 mm.)
Ignition timing	6° B.T.D.C.

IGNITION—continued

Sparking plugs	... Champion NA.8
Plug gap	.025 in. (.635 mm.)

CLUTCH

Type	... Borg and Beck single dry plate
Diameter	9 in. (.25 m.)
Frictional area	36.8 sq. in. (238 sq. cm.) × 2
Withdrawal bearing	Self-lubricating carbon ring
Pedal free movement	$\frac{11}{16}$ in. (17.463 mm.)

GEARBOX

Number of gears	3 forward, 1 reverse
Gear ratios : 1st	2.25
2nd	1.42
3rd	Direct
Reverse	4.981
Oil capacity including overdrive	5½ Imp. pints (6.6 U.S. pints; 3.12 litres)

OVERDRIVE

Make	Laycock-de Normanville
Operation	... Hydraulic
Control	Automatic over-riding switch
Cut-in speed	40 m.p.h. (64 k.p.h.)
Reduction in engine speed : O/D WN1260/1—WN1260/763	32.4%
O/D WN1292/1 onwards	28.6%

PROPELLER SHAFT

Make	Hardy Spicer
Type	Open shaft
Bearings	Needle roller
Universal joints	2

REAR AXLE ...

Type	Spiral bevel
Oil capacity	2¼ Imp. Pints (2.7 U.S. pints; 1.28 litres)
Overall gear ratios : 1st	9.28
2nd	5.85
2nd and O.D. (32.4%)	4.43
2nd and O.D. (28.6%)	4.56
3rd	4.125
3rd and O.D. (32.4%)	3.12
3rd and O.D. (28.6%)	3.28
Reverse	20.5

STEERING

Type	Burman cam and lever
Ratio	... 12.6 to 1
Steering wheel diameter	16½ in. (.42 m.)
Turning circle	35 ft. (10.67 m.)
Front wheel alignment	Toe-in $\frac{1}{16}$ – $\frac{1}{8}$ in. (1.58–3.17 mm.)

SUSPENSION

Front	Independent; coil spring
Rear	... Semi-elliptic leaf spring
Shock absorber	Armstrong double-acting hydraulic
Stabilizer : front	Anti-roll torsion bar
rear	Anti-sway bar

GENERAL SPECIFICATION

BRAKES

Make	Girling
Type	Hydraulic—two leading shoe front
Drum diameter	11 in. (0.28 m.)
Total frictional area	142 sq. in. (916 cm.²)
Handbrake	Mechanical, rear wheels only

WHEELS

Type	Wire spoked
Hub	Knock-on cap

TYRES

Size	5.90 × 15
Pressure: Front (normal)	20 lb./sq. in. (1.406 kg./cm.²)
Rear (normal)	23 lb./sq. in. (1.617 kg./cm.²)

CHASSIS

Type	Integral body and frame
Frame	Box section with cross bracing

ELECTRICAL EQUIPMENT

Battery	2 × 6 volt, positive earth
Capacity	50 amp. hr. at 10 hr. rate
Dynamo	Lucas type C45-PV5
Starter	Lucas type M4198 G
Cut-out and regulator	Lucas type RB. 106
Fuse unit	Lucas type FS6
Heater	Smiths type CHS. 880/11
Windscreen wipers	Lucas

PRINCIPAL DIMENSIONS

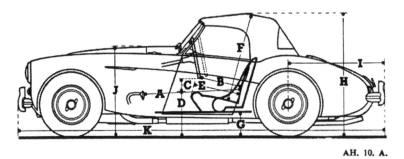

AH. 10. A.

Pedal to seat squab	A	Max. 42¼ in. (1.06 m.) Min. 37¾ in. (0.95 m.)
Steering Wheel to Seat Squab	B	Max. 17¾ in. (0.44 m.) Min. 10¾ in. (0.27 m.)
Steering Wheel to Seat Cushion	C	6 in. (0.15 m.)
Seat to Ground	D	1 ft. 5 in. (0.43 m.)
Seat Cushion Depth	E	1 ft. 7½ in. (0.49 m.)
Seat Cushion to Hood	F	3 ft. (0.91 m.)
Floor to Ground	G	9 in. (0.23 m.)
Overall Height	H	4 ft. 1¼ in. (1.24 m.)
Overhang Rear	I	3 ft. 4 in. (1.02 m.)
Height over Scuttle	J	2 ft. 11 in. (0.89 m.)
Overall Length	K	12 ft. 7 in. (3.83 m.)

PRINCIPAL DIMENSIONS—continued

Overall Width	5 ft. ½ in. (1.52 m.)
Height Over Windscreen	3 ft. 11⅞ in. (1.21 m.)
Wheelbase	7 ft. 6 in. (2.29 m.)
Track front	4 ft. 1 in. (1.24 m.)
Track rear	4 ft. 2¾ in. (1.26 m.)
Ground Clearance	5½ in. (0.19 m.)
Dry Weight (approx.)...	2,176 lb. (987.032 Kg.)

REGULAR ATTENTIONS

THE following is a convenient list of regular attentions which the car should receive to keep it in good mechanical condition. These instructions should be closely followed whether the attentions are performed by the owner or the local agent.

The attentions under the daily or periodic headings are based on the assumption that the maximum mileage per week does not exceed 500, but see "Post Delivery Check" for special attentions during the first 500 miles.

Under more arduous conditions such as very dusty or very muddy roads, long distances at high speeds or heavy roads, it will be advisable to attend to chassis lubrication more frequently.

EVERY DAY

Engine

Check the level of oil in the sump and top up if necessary to the "full" mark on the dipstick. The oil filler is at the rear of the valve rocker cover and the dipstick is on the right-hand side of the engine.

Radiator

Check the level of the water in the radiator and top up if necessary. Fill to just below the filler plug threads when the engine is cold.

Fuel Tank

Check the quantity of fuel in the tank and add upper cylinder lubricant if desired.

Tyres

Check the tyre pressures, when the tyres are cold, not when they have reached their running temperature. See page N/1 for correct pressures.

POST DELIVERY CHECK
First 500 miles (800 km.)

Austin agents are under agreement to carry out a "Post Delivery Check" once during the period of the first 500 miles (800 km.) running, or as soon as possible afterwards, of Austin vehicles purchased from them, when they will, without charge, except for materials used :—

Change the oil in the engine, gearbox and rear axle, and check the oil level in the steering box.

Lubricate all chassis points.

Check the tightness of cylinder head and manifold nuts. Tighten the fan belt if necessary.

Check tappet clearances and ignition timing.

Clean out the carburetter float chamber and check the slow running adjustment. Also clean petrol pump gauze.

Examine and adjust if necessary, the sparking plug and distributor points and verify the working of the automatic ignition control.

Check the front wheel alignment and steering connections.

Check the clutch pedal clearance.

Examine and adjust the braking system.

Check the tightness of nuts and bolts, body and bonnet cowl to chassis, spring clips, etc.

Lubricate the door locks, bolts, hinge pins and seat runners.

Test the lights, check the charging rate, wiring and terminals.

Examine the battery and bring up to the proper level with distilled water or diluted acid.

Test the tyres for correct pressure.

Road test the car after adjustments.

EVERY 1,000 MILES
(1,600 km.)

Gearbox and Overdrive

Check the level and top up if necessary. For access take out the inspection panel in the right-hand side of the gearbox cover when the filler plug will be accessible.

Remove the screw plug and fill up to the bottom of the threads. This gives the correct level.

Rear Axle

Check the oil level and replenish if necessary. The correct grade of oil should be injected into the axle casing from underneath, using the adapter on the oil gun.

First remove the plug, which is on the right-hand lower front side of the axle carrier, then place the end of the adapter into the oil hole, and inject with oil.

The plug also serves as an oil level indicator. Therefore do not replace the plug at once, but give the oil time to run out if too much has been injected. This is most important because if the rear axle is overfilled the lubricant may leak through to the brakes and render them ineffective. Wipe away excess oil from the casing.

Propeller Shaft Splines

Oil the nipple on the sliding yoke at the gearbox end of the propeller shaft. To get at this yoke, lift out the short section of shaft tunnel immediately behind the gearbox cover.

Universal Joints

Lubricate the propeller shaft universal joints. The front joint is best lubricated from above after the short section of tunnel has been removed. The rear joint may be lubricated from below or above through the hinged panel behind the seats. Move the car to bring the nipples to the required positions.

Also test the flange bolts and tighten if these have worked loose; the nuts are secured with tabbed washers.

Steering Connections

Apply the oil gun to the steering centre cross tube nipples (2) and the steering side cross tube nipples (4) and top up the steering idler via the oil plug orifice.

N.B.—On no account should the steering idler be overlooked, as lack of lubricant in this component may cause a serious breakdown due to the additional load imposed on the steering box.

Steering Box

The steering box should be topped up with oil, using the special adapter and the oil gun. Take out the hexagon plug on the side or top (depending on the model) of the steering box to inject the oil. Make certain that grit does not enter the casing during the operation and wipe away any excess oil. To facilitate this attention it is advisable to remove the road wheel.

Steering Column

Lubricate the felt bush at the top of the steering column with a few drops of light machine oil. To gain access to the bush adjust the steering wheel to its full extended position then collapse the spring at the base of the steering wheel hub when the bush will be exposed. Non-adjustable steering wheels have a lubrication hole in the wheel hub.

Swivel Axles

Apply the oil gun to the two nipples on each swivel axle. This is best done when the vehicle is partly jacked up, as the oil is then able to penetrate to the thrust side of the bearings.

Shackle Pins

These are on the rear end of the rear road springs and should be given a charge of oil once a week. There are two nipples, one on each bottom shackle.

Clutch Pedal

With the oil gun, lubricate the nipple at the pivot of the lever.

Brakes

Check the brakes and adjust if necessary (see page M/5). Apply the oil gun to the balance lever on the rear axle, the handbrake pivot, the lubricator on the flexible cable and the pedal pivot nipple.

Brake Fluid Supply Tank

Inspect and refill to the correct level, which is an inch from the top of the container. See Lubrication Chart for recommended fluid.

Brakes and Controls

With the oil can, lubricate all the handbrake linkage points, brake and clutch pedal linkages and carburetter control joints. See Lubrication Chart page S/1.

Shock Absorbers

Ensure that there are no visible signs of leakage and that the rubber bushes are undamaged.

Water Pump

There is a plug on the water pump housing which should be removed and a small charge of oil injected. It is better to under-lubricate than to overdo the attention.

Air Cleaners

Every 5,000 miles the air cleaners should be removed, cleaned and re-oiled. To withdraw the element, first extract the central setpin with its shake-proof washer from the top plate, lift off the plate and pull out the element. Swill the element in petrol, drain, immerse in engine oil and again drain before refitting.

Carburetters

Remove suction chamber cap and damper assembly and replenish oil reservoir as necessary.

Distributor Automatic Advance

Remove the distributor cap and add a few drops of engine oil through the hole in the contact breaker base through which the cam passes.

Distributor Cam-end Drive Shaft Bearings

Lubricate the distributor camshaft bearings by withdrawing the rotor arm from the top of the distributor spindle and carefully adding a few drops of thin machine oil round the screw exposed to view. Take care to refit the rotor arm correctly by pushing it on to the shaft and turning until the key is properly located.

Distributor Cam

Apply a trace of engine oil to the distributor cam. Be careful not to let any oil or dirt reach the contact breaker points.

Battery

Ascertain the state of charge of the two 6 volt batteries by taking hydrometer readings. The specific gravity readings should be:

Fully charged ...	...	1.280–1.300
Half charged ...	...	approx. 1.210
Discharged ...	...	below 1.150

These figures are for an assumed electrolyte temperature of 60°F.

Check that the electrolyte in the cells is level with the tops of the separators. If necessary add a few drops of distilled water. Never use tap water as it contains impurities detrimental to the battery.

Never leave the battery in a discharged condition. If the vehicle is to be out of use for any length of time, have the battery removed and charged about once a fortnight.

Wheels and Tyres

Tighten the wheel nuts and check the tyre pressures, including the spare, using a tyre gauge. Inflate if necessary and see that all valves are fitted with valve caps. Inspect the tyres for injury and remove any flints or nails from the treads. Ensure that there is no oil or grease on the tyres since these substances are harmful to rubber. See Section on "TYRES" for correct pressure.

EVERY 3,000 MILES
(4,800 km.)

Engine

Drain the sump and refill with new oil. Capacity is 11¾ pints (8.71 litres). The oil filter must also be drained and then removed from the engine for refilling. Capacity is 1¼ pints (0.71 litre).

Fan Belt

The fan belt must be sufficiently tight to prevent slip at the dynamo and water pump, yet it should be possible to move it laterally about half an inch each way.

To make any necessary adjustment, slacken the bolts and raise or lower the dynamo until the desired tension of the belt is obtained. Then securely lock the dynamo in position again.

Fuel Pump

The gauze filter of the fuel pump must be removed and cleaned. The filter plug is situated on the under-side of the pump head and must be unscrewed when the gauze strainer and washer will come away with the plug.

Swill the gauze in petrol and lightly blow through with dry air. On replacing the plug and filter ensure that the washer is in good condition.

EVERY 6,000 MILES
(9,600 km.)

Gearbox

Drain when the oil is warm, after a run, and refill to the level of the filler plug with new oil. Capacity, 4½ pints (2.55 litres). The total capacity of the gearbox and Overdrive is 5½ pints (3.12 litres; 6.6 U.S. pints).

After draining, ¼ pint of oil will remain in the overdrive hydraulic system, so that only 5¼ pints will be needed for refilling. If the overdrive has been dismantled the total of 5½ pints will be required.

Overdrive

This oil change is carried out in conjunction with the gearbox attention as the unit has no separate filler plug. However, when draining the gearbox, the Overdrive drain plug, must also be withdrawn. Remove Overdrive oil pump filter and clean the gauze by washing in petrol. The filter is accessible through the drain plug hole and is secured by a central set bolt.

Rear Axle

Drain when the oil is warm, after a run, and refill to the level of the filler plug with new oil. Capacity, 2½ pints (1.42 litres).

Front Road Wheel Hubs

Unscrew the knock-on hub cap and, using the extractor provided, withdraw the grease cap from within the hub. Replace with grease if necessary, but do not over-lubricate as excess grease may penetrate to the brake linings.

The splines and the cone faces of both the hub and wheels, also the threads of the hub cap, should be smeared with grease.

Rear Road Wheel Hubs

These are packed with grease upon assembly and do not require greasing attention other than for wheels, splines and cones as detailed for front wheels and hubs.

Carburetters

The flow of fuel at each carburetter inlet to each float chamber should be checked and if necessary the filters in the union should be cleaned.

To remove a float chamber disconnect the fuel supply pipe, slacken the lid retaining nut and uncouple the steel strut which connects to the induction pipe. Unscrew the chamber holding-up bolt, being careful to note the positions of the brass and fibre washers, and then take off the lid. Do not lose the float needle.

In addition, clean out each suction chamber assembly by removing the three securing screws and lifting off the body in the same plane to avoid damage to the needle.

Lift out the hydraulic damper and wash the assembly in petrol. Dry thoroughly, refit and replenish the damper with oil. When fully re-assembled lift the piston to its fullest extent, thus expelling surplus oil which lubricates the piston rod and eventually finds its way into the induction pipe.

This is the only part which requires lubrication, the piston itself and the inside of the suction chamber should be left dry.

Contact Breaker Points

Clean the contact breaker points. Cleaning of the contacts is made easier if the contact breaker lever carrying the moving contact is removed. To do this, slacken the nut on the terminal post and lift off the spring, which is slotted to facilitate removal. Before replacing smear the pivot on which the contact breaker works with clean engine oil.

Check the contact breaker setting, reset if necessary. The correct gap is .014 to .016 in.

Sparking Plugs

Remove the plugs and clean off all carbon deposit from the electrodes, insulators and plug threads with a stiff brush dipped in paraffin. Alternatively the plugs may be taken to the local Austin dealer for cleaning by specialised equipment.

Clean and dress the plug points and reset to the correct gap of .025 in.

Before replacing the plugs check that the copper washers are in a sound condition. Never over-tighten a plug but ensure that a good joint is made between the plug body, the copper washer, and the cylinder head.

Renewal of sparking plugs is left to the owner's discretion as their efficient working life is variable. Use Champion N.A.8 long reach plugs.

Dynamo Bearings

Unscrew the wick type lubricator cover and if the wick is dry, refill the cup with high melting point grease.

On later models, inject a few drops of oil through the hole in the end caps.

General Check

Examine and, if necessary, tighten all bolts and nuts such as road spring clips, shock absorber retaining bolts, and body panel bolts.

Examine other parts such as steering connections, brake rods, tubing, fuel pipe unions, etc., neglect of any of these points may be followed by an expensive repair and inability to use the car for a lengthy period.

EVERY 9,000 TO 12,000 MILES
(14,400–19,200 km.)

Clutch Operating Shaft

Lubricate the two nipples sparingly, as any excess lubricant may find its way into the clutch.

Speedometer and Tachometer Drives

Disconnect the cables at their instrument end and pull the inner members out of the casings. These should be lubricated sparingly by smearing with light grease. It is important that each drive is *not* over lubricated, otherwise damage will be caused to the instrument should the lubricant find its way into the head.

To reassemble, thread the cable concerned with a twisting movement into the casing, since this will help the cable to engage easily with its union at the gearbox or engine adapter. When this engagement is felt the cable can be pushed home so that the square end stands out approximately $\frac{3}{8}$ in. from the casing. Connect up to instrument head.

SERVICE ATTENTIONS

The following additional inspections and adjustments should be carried out periodically by an Austin dealer at the mileages mentioned. These attentions are not usually carried out by the owner driver and the tools, as supplied in the toolkit, are not sufficient for the work entailed.

EVERY 3,000 MILES
(4,800 km.)

Clutch Pedal Adjustment

Check and adjust if necessary. The pedal should be depressed $\frac{11}{16}$ in. before the clutch springs are felt to be under compression.

EVERY 6,000 MILES
(9,600 km.)

Shock Absorbers

Check the fluid levels and top up if necessary. The correct level is just below the filler plug threads. See page S/2 for recommended fluid. Carefully clear away all road dirt and grit from the vicinity of the filler plugs before removal.

N.B.—Where the recommended fluid is not available the following are acceptable alternatives: Shell Donax A.2, Wakefield's Castrolite, Mobiloil Arctic, Esso Hydraulic (Medium), Duckham's N.P.20, B.P. Energol S.A.E.20.

Track Alignment

Check front wheel alignment, $\frac{1}{16}$ to $\frac{1}{8}$ in. toe-in taken along a horizontal line at centre height using the wheel rims as data points.

Ignition Timing

Check setting and adjust if necessary.

EVERY 9,000 TO 12,000 MILES
(14,400–19,200 km.)

Steering Box

Check for wear. This can be felt if the front wheels can be moved without creating any movement at the steering wheel.

Front and Rear Hub Bearings

Check for signs of wear.

Starter and Dynamo Commutators

Clean, also check freedom of brushes in holder.

External Oil Filter

Undo the "Full Flow" filter central fixing bolt and empty the container. Take out the old element, and replace with a new one. Tecalemit type FG.2313 or Purolator type M.F.26A, whichever is applicable.

Capacity of the filter is approximately $1\frac{1}{4}$ pints; prime before refitting.

TOP OVERHAUL

Decarbonising, Valve Grinding and Adjustment

This attention may not be needed so frequently on cars used for long journeys. As a general guide, a falling off in engine power indicates when decarbonising is due. The owner is advised to take his car to the local Austin Agent for examination. The correct valve clearance is .012 in. with the engine hot or cold.

EVERY 30,000 MILES
(48,000 km.)

Oil Sump

Remove and clean sump and oil pump strainer gauze.

OIL GUN

The gun, as supplied, is used for forcing lubricant through the nipples. Charge the gun by unscrewing the end cap and fill to its capacity.

Oiling Technique

Always make sure that the nipple on the chassis component, about to be lubricated, is clean before applying the gun. Push the gun body hard and repeat the strokes according to the amount of lubricant required in the component. Whenever possible watch for old oil exuding from the component concerned, as this is proof that the new oil is being forced in. A nipple that refuses to pass oil should be removed and cleaned. This is best achieved by leaving the nipple to soak for a short time in paraffin.

Should difficulty be experienced in the operation of the gun it is probably due to air locks. This can be easily overcome by carrying out the following procedure:—

Extend the steel cylinder as far as possible, fill the gun with the correct oil and replace the cap, hold the gun firmly in the left hand, unscrew the cap approximately two turns and then gently force the steel cylinder into the gun. This will force the oil to the top of the barrel and displace any air that may have been included in the filling process; the air can be heard distinctly escaping past the threads of the cap and when the oil begins to emerge, the cap should then be tightened. After lubricating a point, it is essential that the disconnecting process should be made with a sideways breaking movement and not pulled directly away; any attempt to disconnect the gun by pulling directly will have a tendency to break the spring clip in the nozzle of the gun and at the same time to extend the cylinder, thereby sucking in air.

There is an adapter supplied with the oil gun to facilitate the topping up of such components as the rear axle and steering box. The procedure is as follows: Remove the end cap and extend the steel cylinder as far as possible, fill the gun with the recommended oil and then screw on the adapter in place of the end cap.

Remove the steering box or rear axle filler plug, insert the adapter end into the filler orifice and force the steel cylinder into the gun body. This will quickly empty the gun's contents into the component concerned.

Replace the filler plug after ascertaining that the component has been topped up to its correct level.

INSTRUMENTS AND CONTROLS

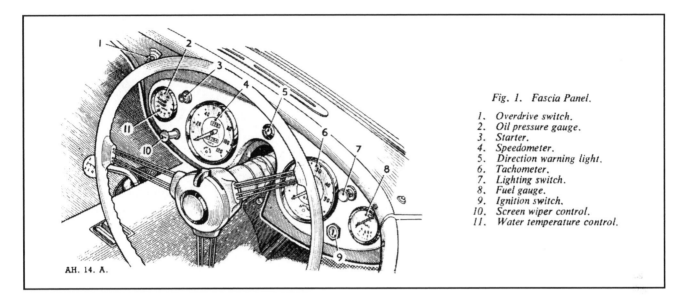

Fig. 1. Fascia Panel.

1. Overdrive switch.
2. Oil pressure gauge.
3. Starter.
4. Speedometer.
5. Direction warning light.
6. Tachometer.
7. Lighting switch.
8. Fuel gauge.
9. Ignition switch.
10. Screen wiper control.
11. Water temperature control.

AH. 14. A.

INSTRUMENTS

Speedometer

Registers the car speed and total mileage. The trip figures at the top of the speedometer can be set to zero by pushing up the knob at the bottom of the speedometer and turning it to the right.

Tachometer

This instrument indicates the revolutions per minute of the engine and thus assists the driver in determining the most effective engine speed range for maximum performance in any gear.

Oil Pressure Gauge

Indicates the oil pressure in the engine. It does not show the quantity of oil in the sump.

Ignition Warning Light

Glows red when the ignition is switched "on" and fades out when the dynamo is charging the battery.

Headlight Beam Warning Light

A red light appears when both headlights are switched on, with the two beams full ahead. The light goes out when the headlights are dipped.

Direction Flasher Warning Light

A green light starts to blink as soon as the direction indicator switch is operated. It is fitted to warn the driver that the indicator lights have been left flashing, as they cannot be seen from the driver's seat.

Fuel Gauge

Indicates the contents of the petrol tank when the ignition is switched on. When the tank is being filled switch off to stop the engine, switch on again and the needle will record the amount of fuel entering the tank.

Water Temperature Gauge

This records the temperature of cooling water circulating in the cylinder block and radiator. The correct running temperature under normal conditions should be 185°–194°F. approximately.

FOOT CONTROLS

Accelerator

"Organ" type pedal at right-hand side.

Brakes

The centre pedal operates the hydraulic brakes on all four wheels.

Clutch

The left pedal. Do not rest the foot on this pedal when driving and do not hold the clutch out to "free wheel."

Dip Switch

If the headlights are on full, a touch on the foot dip switch alters the lights to the "dipped" position where they remain until another touch returns them to "full-on."

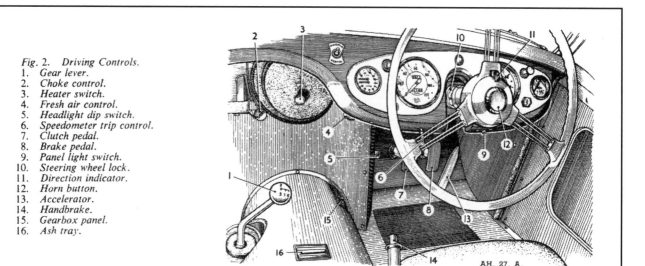

Fig. 2. Driving Controls.
1. Gear lever.
2. Choke control.
3. Heater switch.
4. Fresh air control.
5. Headlight dip switch.
6. Speedometer trip control.
7. Clutch pedal.
8. Brake pedal.
9. Panel light switch.
10. Steering wheel lock.
11. Direction indicator.
12. Horn button.
13. Accelerator.
14. Handbrake.
15. Gearbox panel.
16. Ash tray.

HAND CONTROLS

Handbrake

Situated on the right-hand side of the propeller shaft tunnel, between the seats. To release, pull rearwards slightly and depress button, then push forward fully. Operates on the rear wheels only.

Gear Lever

Should always be left in neutral when starting the engine. The lever is mounted on the left side of the gearbox. To engage a gear, depress the clutch and move the lever to the required position.

Choke Control

Located on the left side of the heater, underneath the fascia. For use when the engine is cold. Pull out to the limit until the engine fires and return it to the half-way position for rapid warming up. The choke must be released at the earliest moment possible.

Ignition Switch

Turn the key clockwise to switch on. Do not leave the switch "on" when the engine is stationary; the warning light is a reminder. The ignition key may also be used for locking the luggage compartment.

Lighting Switch

·Mounted above the ignition switch. Pull out to put on sidelights; twist to right and pull again to put on headlights. The headlights are dipped by foot operation.

Starter Switch

Press in the control to start and release as soon as the engine fires. If the engine fails to start after a few revolutions, do not operate the starter again until the engine is stationary.

Direction Flasher Switch

The indicator lights are controlled from the centre of the steering wheel. Normally, after the car has turned a corner, the control is automatically returned to the vertical position and the lights switched off, but when only a slight turn has been made it may be necessary to return the switch manually.

Windscreen Wipers

To start the electric wipers, gently pull out the wiper control. To park, switch off by pressing the control inwards when the arms are at the end of their stroke. Do not try to push the arms across the screen by hand.

Panel Light Switch

Slide the switch to the right to illuminate the instruments. Only operates when the side lights are "on".

Overdrive Switch

A chromium plated switch is mounted centrally on the fascia panel with two positions clearly marked. When the switch is moved to the overdrive position, the overdrive, which is governed, will automatically come into operation for speeds of 40 m.p.h. and over.

Heater Switch

Mounted on the heater unit which is situated below the centre of the fascia. To start, turn the switch to the right until a click is heard. The further the control is turned the less will be the speed of the fan, due to the fact that a rheostat is incorporated in the switch.

Battery Master Switch

This switch, situated in the luggage compartment recess, adjacent to the spare wheel, is fitted as an anti-thief device. The luggage compartment must of course be locked after the switch has been turned to the "off" position.

Fig. 3. After parking, the car may be made 'safe' by operating the battery master switch, then locking the luggage compartment.

Air Intake Control

A supply of cold air, entirely independent of the heater unit, can be admitted to the car interior for ventilating purposes by pulling out the control on the right-hand side of the heater and removing the rubber grommet on the underside of the scuttle valance (right-hand side).

Horn Button

Mounted at the centre of the steering wheel operated independently of the ignition switch.

Steering Column Adjustment (Early Models)

The steering column may be adjusted to give the most suitable steering wheel position. To raise or lower the steering wheel, first release the locking ring immediately behind the steering wheel hub. Slacken this locking ring two or three turns in a clockwise direction, move the steering wheel up or down as desired then re-lock in the new position.

Windscreen

To gain the high speed position unscrew the knurled knob of each pillar to free the locating pegs and to clear the windscreen wipers, then move the windscreen

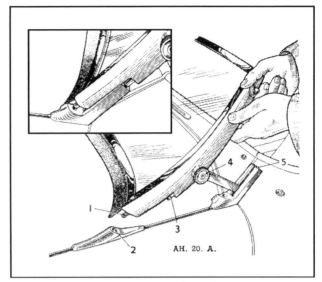

Fig. 4. Adjusting windscreen for high speed position.
1. *Locating peg.*   2. *Position for (1).*   3. *Locating peg.*
4. *Knurled locknut.*   5. *Location for (3).*
Inset shows upright position for screen.

forwards so that the peg on each pillar fits into its forward locating hole. Finally, tighten the knurled knobs.

Sidescreens

Sidescreens can be speedily erected or removed from the doors. The bottom rail of each sidescreen has two pegs fitted to it and these pegs locate into wells sunk into the top edge of each door. Early type sidescreens were of one piece design as illustrated in fig. 5 whilst the later type have hinged flaps incorporated in order that hand signals can be given.

Fig. 5. Fitting a sidescreen of the early type.

Hood

To stow away the hood, first release the securing locks at each windscreen pillar. The hood will then spring away from the screen. Next undo the turn buttons and stud fasteners at the rear and side of the hood, and inside the door rear pillar, then pull the steel bar of the hood rear panel out of the two chrome clips on the boot top panel.

Bring the two main ribs of the hood frame together, folding the material inwards between them. Close the trellis frames at each side of the hood so that the front rail moves rearwards to meet the first rib. Again fold the material neatly between rail and rib. With the collapsed hood upright ensure that the rear panel hangs straight down, then swing the whole assembly forward and downward to tuck away behind the seats.

Raising the hood is an exact reversal of this procedure.

Bonnet Catch

To open the bonnet pull the control handle mounted centrally behind the fascia. The rear edge of the bonnet will rise an inch or so to be held by a spring loaded safety catch. Push back the catch and lift the bonnet, which can be held open by a rod, hinged to the bonnet surround and locating in the bonnet top.

When closing the bonnet, exert a slight downward pressure on the bonnet top until the locking catch is heard to engage.

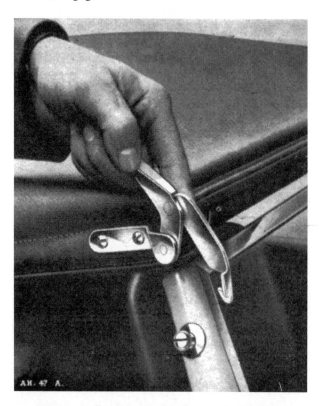

Fig. 6. An illustration showing the operation of releasing the hood securing hooks at the windscreen pillars.

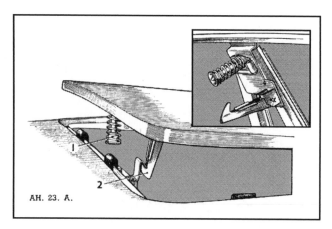

AH. 23. A.

Fig. 7. When the bonnet control is operated the spring (1) will raise the bonnet until it is caught in position by the safety catch (2). Inset shows catch and spring enlarged.

DRIVING INSTRUCTIONS

Running In

It is most important to remember that at no time during the first 500 miles running-in period, must the engine be overloaded such as attempting to ascend steep inclines in top gear at low vehicle speeds. The load should be eased by changing down to a lower gear.

Fierce acceleration must also be avoided, and remember that the engine should never be raced in neutral.

On completion of the first 500 miles the running speed in each gear may be progressively increased, but full power should not be used until at least 1,500 miles have been covered, and even then only for short periods of time.

During this period a slight falling off in engine power may develop, in which case it will be beneficial to lightly grind in the valves and reset the valve clearance. No engine or complete car can be considered fully run-in until it achieves 2-3,000 miles.

The use of upper cylinder lubricant is advocated at all times, but most particularly during the first 2,000 miles. See Section 'S' for recommended brands.

Running-In Speeds

3,000 r.p.m. for the first 500 miles. 4,000 r.p.m. for the first 500–1,000 miles.

Starting

Before starting the engine, see that the gear lever is in neutral and that the handbrake is applied.

Switch on the ignition and press the starter control firmly. Do not continue to press the starter button if the engine fails to start promptly. Allow a short interval between each successive attempt to start, and if the engine does not start in a reasonably short time, look for the cause of the trouble. Never press the starter button unless the engine is stationary.

As soon as the engine starts, release the starter button and push in the choke control to the halfway position. Release the choke completely as soon as the engine will run without it.

Do not allow the engine to race when first starting up as time must be allowed for the air to circulate properly. Let the engine idle fairly fast for a few minutes before moving off, and engage top gear as soon as possible. Blanketing of the radiator will assist the engine to warm up quickly in cold weather, but always uncover the radiator before driving off. There is a thermostat to assist rapid warming up.

Driving

The gearbox has three forward speeds and reverse and incorporates a Laycock-de Normanville overdrive unit which gives two additional forward gear ratios. Start only in first gear, which is engaged by depressing the clutch pedal and moving the gear lever a little to the right and rearwards as indicated on top of the gear lever (see fig. 2). Gradually release the clutch pedal, at the same time gently depressing the accelerator and releasing the handbrake. The car will move forward, gathering speed in accordance with the amount the accelerator is depressed.

Second gear is engaged by depressing the clutch pedal, moving the gear lever forwards, then to the left and forwards again. Release the clutch pedal. Ease the accelerator when the higher gear is engaged.

To engage high gear (top), move the gear lever rearwards as far as it will go.

Changing down is effected by reversing the above procedure, with the exception that the accelerator pedal should be kept depressed whilst changing gear, in order to speed up the engine to suit the lower gear speed.

To engage reverse, which must only be done when the car is stationary, move the gear lever hard over to the right and rearwards. Remember, however, that the gearing is now lower than first gear. Therefore, release the clutch very slowly until the car just begins to move and then gently depress the accelerator to give the desired speed.

When temporarily halted on an incline do *not* slip the clutch—use the handbrake.

When descending a steep hill, it is advisable to engage a low gear as the engine will then provide a useful braking action.

Overdrive

To engage overdrive the driver must first move the fascia panel switch to the overdrive position. Normal drive is then in engagement up to a speed of approximately 40 m.p.h. at which speed an automatic governor engages overdrive.

If the car speed is reduced after having engaged overdrive the automatic governor is set such that normal

drive would not be re-engaged until the speed drops to approximately 30 m.p.h. However, although the governor makes it possible to automatically engage normal drive below 30 m.p.h., the engagement does not actually occur until the accelerator pedal is depressed an amount sufficient to operate the throttle switch. This switch is normally set to operate at a very small throttle opening so that direct drive is engaged as soon as the pedal is depressed for accelerating the car.

Should the overdrive be in engagement at a speed above 30 m.p.h. when direct drive is required, then the fascia switch can simply be moved to normal drive position when the normal gear will be automatically engaged.

The fascia switch should not be moved into the normal position at speeds which are in excess of the direct drive maximum (see following table).

If the car is brought to rest without the accelerator pedal being depressed to disengage the overdrive, then upon starting off, in either first or reverse gear, a special switch incorporated in the gearbox and operated by the gear lever re-engages normal drive. However, if starting from rest in second gear normal drive is engaged as soon as the throttle switch is contacted.

The following table gives the relationship between engine revolutions per minute to read speed in miles and kilometres per hour for the various gears. The top and second gear columns are divided to show the comparative engine revolutions with and without overdrive in operation.

Road Speed		Engine R.P.M.				
K.P.H.	M.P.H.	1st	2nd Norm.	2nd +OD.	Top Norm.	Top +OD.
16	10	1250	790	—	558	—
32	20	2500	1580	—	1115	—
48	30	3750	2380	—	1675	—
64	40	5000	3160	2400	2230	1685
80	50		3900	3000	2790	2110
96	60		4750	3600	3350	2535
112	70			4200	3910	2955
128	80			4800	4460	3375
144	90				5000	3800
160	100					4220
176	110					4640

Driving Hints

Do not press the starter control when a gear is engaged.

Remember to switch on the ignition before attempting starting the engine.

Refrain from continual pressing of the starter control if the engine will not fire.

Release the choke control as soon as possible after starting the engine.

Avoid leaving the car in gear with the handbrake off.

Never engage reverse gear when the vehicle is moving forward or a forward gear when the vehicle is moving backwards. Serious damage may result.

Do not slip the clutch in traffic or on an incline.

It is bad driving, apart from being injurious to the clutch, to coast the car with a gear engaged and the clutch pedal depressed.

Refrain from running the engine at high speeds for the first 500 miles.

Abstain from racing the engine in neutral at any time.

Remember not to run the car with the radiator completely blanked off.

Never fill the radiator with cold water when the engine is hot.

Do not under any circumstances run the engine in a closed garage or similar restricted atmosphere. The exhaust fumes are highly poisonous and if inhaled will quickly produce grave, if not fatal results.

COOLING SYSTEM

A N efficient cooling system is of major importance to ensure the satisfactory running of the engine and it is therefore necessary to pay particular attention to its maintenance. Attention is especially drawn to the procedure advised for the winter months, if damage is to be avoided.

Description

The cooling system is maintained by water pump circulation combined with an efficient fan-cooled radiator and thermostat.

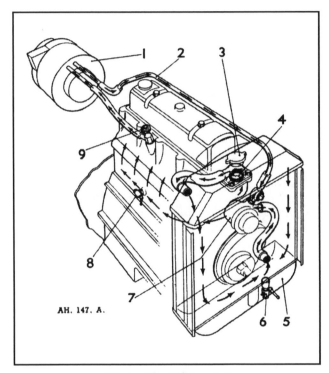

Fig. 1. The cooling system.
1. Heater.   2. Return pipe.   3. Filler cap.   4. Thermostat.
5. Radiator bottom tank. 6. Radiator drain tap. 7. Water pump.
8. Cylinder block drain tap. 9. Control valve.

The system is pressurised and the relief valve, incorporated in the radiator filler cap, controls the pressure at 7 lb. per sq. in. N.B.—Do not remove the filler cap if the temperature of the coolant is above boiling point or if the engine is running.

Topping-up should only be required occasionally to replace water lost through the overflow pipe. Top-up when the engine is cold, and if possible use rain water or clean soft water. Fill up to just below the filler cap orifice. The capacity of the system is 20 pints.

Thermostat

In order to ensure maximum efficiency, it is essential to keep the engine operating temperatures within certain limits. To assist this, a bellows type thermostat is fitted,

being located in the water outlet at the front of the cylinder head. The device consists of metallic bellows, filled with a volatile liquid, which control a mushroom valve.

When the engine is cold this valve is closed and on starting the engine the flow of water to the radiator is temporarily restricted. Due to this, the temperature of the water in the cylinder head and cylinder jackets will quickly rise, thus ensuring rapid warming up. The heat so generated will gradually expand the bellows so opening the valve, and ultimately permitting a full flow of water to the radiator.

The thermostat itself is detachable; therefore, should occasion arise, it can be removed from its housing and the hose reconnected to avoid laying up the car. Should the thermostat be tight there are two tapped holes on the top which may be utilised to ease it from the casting.

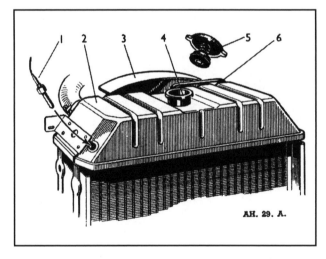

Fig. 2.   1. Thermometer bulb.   2. Radiator top tank.   3. Fan guard.   4. Filler orifice.   5. Filler cap.   6. Overflow pipe.

When the system has been completely emptied, it is essential to wait a minute or two after refilling to allow air to escape through the thermostat valve and then finally top up.

The thermostat opening is set by the manufacturer and cannot be altered. It opens at a temperature of 70°-75°C. During decarbonising it is policy to test this opening by immersing the thermostat in water raised to the requisite temperature. The valve should open under these conditions, but if it fails to open a new unit should be fitted.

Overheating

Overheating may be caused by a slack fan belt, excessive carbon deposit in the cylinders, running with ignition too far retarded, incorrect carburetter adjustment, failure of the water to circulate or loss of water.

Fan Belt Adjustment

The fan is driven from the crankshaft by a "V" belt, this also driving the dynamo. A new belt can be fitted by first loosening the clamp bolts (Fig. 3), which hold the dynamo in position and moving the dynamo towards the engine. Slide the belt over the fan and on to the fan pulley, the crankshaft pulley, and finally on to the dynamo pulley. Adjustment is then made by bringing the dynamo away from the engine. The belt should be sufficiently tight to prevent slip, yet it should be possible to move the belt laterally about one inch

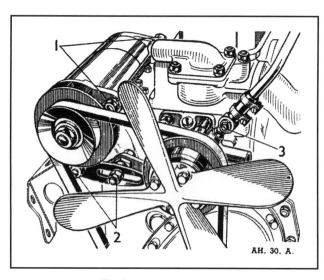

AH. 30. A.

Fig. 3.   Fan belt adjustment.
1. *Dynamo fixing bolts.*   2. *Fan belt adjusting nuts.*   3. *Heater return pipe cock.*

each way. As the drive is taken on the "V" of the pulleys it is not necessary to have the fan belt tight; to do so may cause excessive wear to the dynamo and water pump bearings. After the correct tension has been obtained, securely lock the dynamo in position again.

Frost Precautions

Freezing may occur first at the bottom of the radiator or in the lower hose connections. Ice in the hose will stop water circulation and may cause boiling.

A muff can be used to advantage, but care must be taken not to run the car with the muff fully closed, or boiling will result.

Anti-freeze Solution: Cars with anti-freeze mixture in the cooling system should have an identification mark on the header tank of the radiator, under the bonnet, in the form of a disc painted in a special colour.

The following precautions are necessary on cars so marked :—

1. When frost is expected or when the car is to be used in a very low temperature, make sure that the strength of the solution is, in fact, up to the strength advised by the manufacturers.
2. The strength of the solution must be maintained by topping up with anti-freeze solution as necessary. Excessive topping up with water reduces the degree of protection afforded. Solution must be made up in accordance with the instructions supplied with the container.

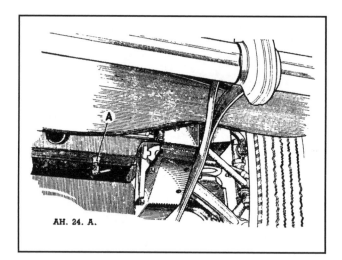

AH. 24. A.

Fig. 4.   Showing the radiator drain tap A in the open position. Turn the tap lever down to close.

3. **Top up when the system is cold.**
4. If the cooling mixture has to be drained, run the mixture into a clean container and use again.
5. If for any reason the mixture is lost and the system is filled with water only, **remove the painted disc on the header tank.**

Protection by Draining: On cars where anti-freeze is not used the following precautions must be taken during frosty weather to obviate any damage due to freezing of the cooling system.

When heavy frost is imminent, the cooling system must be completely drained. It is not sufficient merely to cover the radiator and engine with rugs and muffs. **There are two drain taps, one on the left-hand side of the cylinder block and the other at the base of the radiator block. Both taps must be opened to drain the system and the car must be on level ground while draining.** The drain taps should be tested at frequent intervals by inserting a piece of wire to ensure that they are clear. This should be done immediately the taps are opened, so that any obstruction freed by the wire may be flushed out by the water. The draining should be carried out when the engine is hot.

When completely drained the engine should be run for a timed minute to ensure that all water has been cleared from the system. A suitable notice should then be affixed to the radiator, indicating that the water has been drained. As an alternative, place the radiator filler cap on the driver's seat or leave the bonnet unlocked as a reminder to fill the cooling system before the car is used again.

N.B.—If a heater is fitted, under no circumstances should draining of the cooling system be resorted to as an alternative to the use of anti-freeze, due to the fact that complete draining of the heater unit, by means of the drain taps, is not possible.

Flushing the Radiator

To ensure efficient circulation of the coolant and to reduce the formation of scale and sediment in the radiator, the system should be periodically flushed with clean running water, preferably before putting in anti-freeze in the Autumn and again when taking it out in the Spring.

The water should be allowed to run through until it comes out clean from the drain taps.

Fig. 5. *Showing the cylinder block drain tap in the closed position. Turn the tap lever up to open.*

At intervals a stiff piece of wire should be inserted into the taps during draining to ensure that they are not becoming clogged with sediment.

This method of radiator flushing may serve well, but in cases where the "furring" up is excessive the

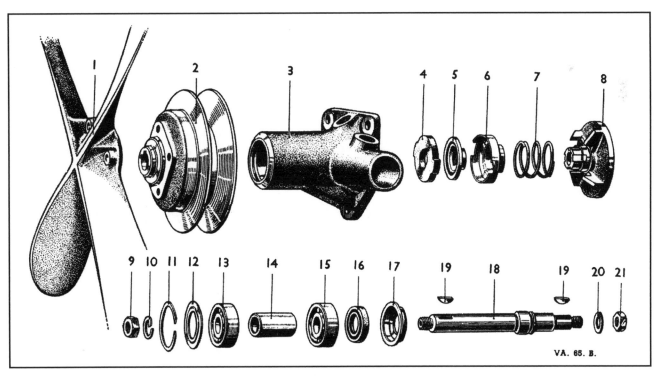

Fig. 6. *Water pump assembly.*

1. Fan.	8. Water impeller.	15. Ball race.
2. Pulley.	9. Spindle nut.	16. Rubber seal.
3. Pump body.	10. Washer.	17. Seal cap.
4. Carbon ring.	11. Split ring.	18. Pump spindle.
5. Rubber seal.	12. Thrust washer.	19. Woodruff keys.
6. Seal housing.	13. Ball race.	20. Washer.
7. Spring.	14. Distance piece.	21. Spindle nut.

operator will find it a more efficient practice to remove the radiator completely and flush in the reverse way to the flow, i.e., turn the radiator upside down and let water flow in through the bottom hose connection and out of the top connection.

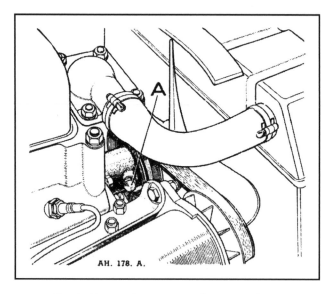

Fig. 7. Showing the water pump lubrication plug A.

WATER PUMP AND FAN

Removing and Dismantling

Drain the water from the radiator. Remove the pump unit from the cylinder block by taking the nuts and washers from the three studs in front of the block, disconnecting the lower hose and the interior heater return pipe.

If necessary, remove the four-bladed fan by withdrawing four setpins from the pulley.

Removing the Pump Spindle

Remove the nut from the front end of the spindle, withdraw the fan pulley and take out the key.

While holding the pump body, the spindle can now be tapped out towards the rear, carrying with it the impeller, spring and washers.

Bearings and Washers

The ball bearings, distance piece, steel and rubber washers have next to be removed from the body. First prise out the spring retaining ring and remove the oil retaining ring; then, using the water pump bearing drift (Service Tool 18G 61), tap out the first bearing, which will be followed by the tubular distance piece. The second bearing must be centralised in the body before it can be tapped out. It will be followed by the oil retaining assembly, consisting of a dished steel washer and a rubber seal.

Sealing Ring Assembly

Removing the nut on the rear end of the spindle, will enable the impeller and key to be withdrawn, followed by the spring metal cup washer, rubber washer and carbon sealing ring. The latter registers within the cup washer.

Reassembly

When reassembling, it is essential that the bearings, distance piece, and various washers, together with other parts, be positioned correctly; Fig. 6 shows the correct order.

Lubrication

Lubricate sparingly with oil, using the oil gun after removing the plug, see Fig. 7.

Replacing the Pump

When refitting it is most important to ascertain that the gland spring is holding the carbon seal against the pump body at the correct pressure.

This can be done by making sure that the gland spring is just holding the carbon seal up against the shoulder on the spindle before inserting it in the pump body.

REMOVING THE RADIATOR

The radiator is held in position by two setpins and two bolts, one of each at either side.

To remove the radiator, first drain the water from the system then, from beneath the bonnet, release the upper water hose from the thermostat housing. To effect this removal, slacken the hose clip with the aid of a screwdriver then ease the pipe off the housing extension. In a similar manner release the lower hose

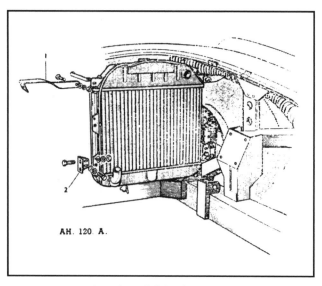

Fig. 8. Showing the radiator left-hand mounting points exploded at 1 and 2. The right-hand mounting points are identical.

from the water pump. The temperature gauge bulb must be removed from the header tank.

Fig. 9. The fresh air control valve.
1. Butterfly valve. 2. Control lever.

Lift out the radiator after releasing the four securing points, see Fig. 8. This operation will be made easier if the bonnet top is removed. See Bodywork Section.

When replacing the radiator reverse the removal procedure, then fill up the system with water or the requisite mixture of water and anti-freeze.

HEATER UNIT

The heater is mounted below the centre of the fascia and is secured by three nuts and washers to the driver's side of the engine compartment rear bulkhead. Two of these points are accessible on this side, immediately behind the heater; the other from the engine side of the bulkhead when the rubber grommet, situated below the bonnet locking mechanism, is removed.

The mechanism of the heater unit consists of a radiator and an electric motor-driven fan. Water from the engine cooling system passes through the heater radiator, thus warming the air drawn into the heater by the fan. The latter also serves to dispose the warm air to the car interior via a grille in the forward part of the heater. The air for demisting is supplied to the windscreen by two flexible pipes, having a common junction box at the heater.

In the summer period, the flow of hot water may be completely shut off and instead, fresh air drawn in by operating the air control (see pages A/2, A/3) and removing the large rubber grommet beneath the scuttle valance on the right-hand side.

Water from the cylinder head is fed to the heater via a control valve. A return pipe feeds the water back to the engine water circulating pump. When shutting off the supply of hot water to the heater, in addition to closing the cylinder head valve (see Fig. 1), there is a shut-off cock in the return pipe to be closed. This cock is situated at the junction of the return pipe and water pump. The object of shutting off this cock is to ensure that heat is not transferred back to the heater unit.

Removal

To remove the heater, first drain the cooling system and then release the two hose clips from the pipes protruding forward and out of the heater unit through the bulkhead. On the other side of the bulkhead release the two electrical leads from the heater motor, one from its snap connection and the other from the bulkhead by undoing a nut.

Fig. 10. The heater unit.
1. Heater motor cables.   2. Water outlet pipe.
3. Water inlet pipe.   4. Demister pipes.
5. Switch.

Before finally releasing the main securing points (previously mentioned) and removing the heater, withdraw the push fit de-mister pipe junction box from the heater. If desired, the demister flexible pipes can be removed when the clips that secure them behind the fascia are released.

The heater motor does not normally require servicing, as the bearings are packed with lubricant on assembly.

SERVICE DIAGNOSIS GUIDE

Symptom	No.	Possible Fault
(a) Internal Water Leakage	1	Cracked cylinder wall
	2	Loose cylinder head nuts
	3	Cracked cylinder head
	4	Faulty gasket
	5	Cracked tappet chest wall
(b) Poor Circulation	1	Radiator core blockage
	2	Water jacket restriction
	3	Low water level
	4	Loose fan belt
	5	Defective thermostat
	6	Perished or collapsed radiator bottom hose
(c) Corrosion	1	Impurities in water
	2	Infrequent draining and flushing
(d) Overheating		In (b) check 4, 5 and 6
	1	Sludge in crankcase
	2	Faulty ignition timing
	3	Low oil level in sump
	4	Tight engine
	5	Choked exhaust system
	6	Binding brakes
	7	Slipping clutch
	8	Valve timing incorrect
	9	Retarded ignition
	10	Mixture too weak

FUEL SYSTEM

ALTHOUGH the fuel system of the Austin-Healey "100" is described here in detail, any measure of intricate servicing is best left to the expert.

From the rear fuel tank, capacity 12 Imperial gallons, an S.U. electric fuel pump transfers the fuel from the tank to the twin S.U. Carburetters each of which has its own Burgess air cleaner-silencer.

PETROL TANK

Tank Draining

A screwed plug is provided in the base of the fuel tank, at the front and to the left, in order that the tank may be drained of fuel, should the occasion arise, upon removal of this hexagon-headed plug.

Tank Removal and Replacement

First drain the tank then disconnect the petrol delivery pipe from its tank union which is situated at the front left-hand side of the top face of the tank. By opening the luggage compartment lid access to this union can be gained.

Take out the carpet covering of the tank then disconnect the insulated lead from the petrol gauge unit terminal situated at the front left of the tank top.

Finally release the tank securing straps by undoing the nut and locknut of each tank strap stud. These nuts are visible on the underside of the luggage compartment floor just in front of the rear body panel. Pull the straps through the compartment floor and hinge them back on their clevis pin anchorages placed upon the front bulkhead of the compartment. Lift out the tank.

Replacing of the tank is simply a reversal of the removal operations.

Tank Gauge Unit

This can be removed from the tank complete by the withdrawal of the six securing screws, but care must be

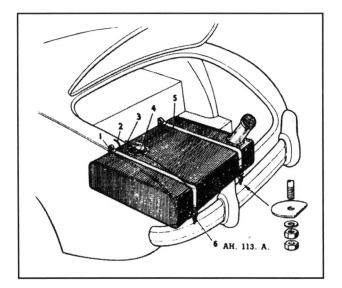

Fig. 1.   Illustrating the position of the fuel tank in the luggage compartment— 1. Strap clevis pin anchorage. 2. Fuel delivery pipe. 3. Lead to tank gauge unit. 4. Tank gauge unit. 5. Securing strap. 6. Strap nut and locknut.

taken not to bend or strain the float lever otherwise subsequent gauge readings may be seriously affected.

Great care should be taken, when refitting the gauge unit to see that the joint washer is in place. It is essential that a petrol tight joint should be made between the tank and the face of the unit. If there is any apparent damage to the washer it must be replaced by a new one.

MANIFOLDS AND EXHAUST SYSTEM

To remove both induction and exhaust manifolds it is first necessary to drain the cooling system as the heater return pipe must be removed during the dismantling operations.

After releasing the heater return pipe hose clips at both ends, undo the nuts of the two studs protruding upwards from the inlet manifold balance pipe and lift the heater pipe complete with clips from the balance pipe studs.

Take off the outer case and element of each air cleaner-silencer leaving the back plates of the filters attached to the carburetter flanges. Each outer case and element is held in place by a single setpin through the top plate.

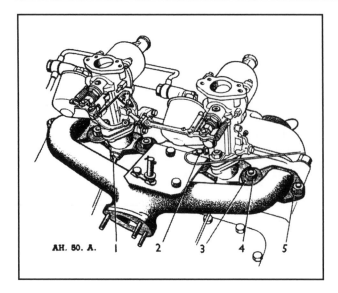

Fig. 2. Manifold assembly — 1. Induction manifold drain. 2. Vacuum pipe for ignition advance. 3. Manifold clamping washers. 4. Manifold inner nuts. 5. Manifold outer brass nuts.

care at this juncture not to lose or damage the springs and thimble filters or washers at each float chamber head.

To release the choke cable, first slacken the screw securing the inner cable. This screw is situated below the carburetter bridging stay and immediately above the wire that links the two choke operating arms. Release the outer cable at the bridge stay by slackening the securing nut and clamp bolt. Withdraw upwards the choke cable complete and tuck it away out of the working area.

Next lift off each carburetter supporting stay, having first released them at the head of each float chamber where they are secured by the central brass cap nuts. At their inner ends these stays are held by the balance pipe studs previously mentioned.

Disconnect the carburetter throttle rod from its ball joint connector at the manifold bell crank lever, also disconnect, in the same manner, the rod that links the manifold and scuttle bell crank levers together. It is best to make this disconnection at the scuttle bell crank.

Each carburetter is secured to its short induction manifold by two flange nuts and bolts. Undo these nuts and withdraw the bolts when the carburetters, still connected to one another by the common throttle rod, can be separated from the manifolds as one unit.

At the lower end of the exhaust manifold the exhaust downpipe is secured by three flange studs and nuts. Undo these nuts, also slacken the "U" bolt nuts that steady the pipe by means of a crankcase bracket. Drop the downpipe away from the manifold studs.

Disconnect the breather pipe from both the front air cleaner and the rocker cover.

One common fuel feed pipe feeds both carburetters. This pipe must be relieved of the union at its lower end where it joins up with the flexible fuel pipe. Release the pipe clip at the engine front mounting plate by extracting the clip securing setpin.

Also unscrew each banjo union at its float chamber connection and remove the fuel pipe complete. Take

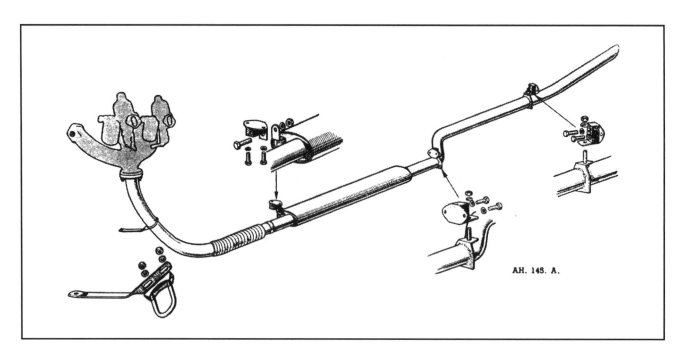

Fig. 3. This illustration shows the exhaust system assembled, whilst the enlargements illustrate the fixing of the system to the car frame and crankcase. The exhaust system is in three sections: downpipe, silencer and tailpipe.

There is now working clearance to extract the two tappet cover setpins which secure the induction manifold drain pipe clips.

Finally undo the six brass nuts that hold the manifolds in place. The four inner nuts, it will be observed, secure both induction and exhaust manifolds via special clamping washers and spring washers. Withdraw the induction manifolds first then the exhaust manifolds.

Reassembly of the manifold units is a reversal of the removal procedure.

Exhaust Pipe

To dismantle a complete exhaust pipe, the first operation is the disconnection of the downpipe from three manifold flange studs. If the carburetters, induction and exhaust manifolds are to remain secured to the cylinder head, the downpipe securing nuts can be reached from beneath the car.

The downpipe "U" bolt, already detailed under "Manifolds", must be released, also slacken the bracket at its crankcase end.

Moving towards the rear of the car it will be found that there is a bracket attached to the front end of the silencer which secures the exhaust system to the chassis frame. Extract the single bolt fixing the bracket to the silencer.

There are two further brackets securing the tail pipe to the chassis frame. Of these brackets, which are identical, one is fixed to the rear cross member of the frame whilst the other is placed to the rear of the silencer. A single nut and stud holds each bracket.

With these points released the complete exhaust pipe can be lowered and removed from the car. However, should it only be necessary to remove the silencer then the two tail pipe brackets must be released followed by the slackening of the two circular clip bolts that secure one end of the silencer to the tail pipe and the other end to the downpipe. Pull the tail pipe to the rear of the car and clear of the silencer then separate the silencer from the downpipe.

THROTTLE LINKAGE

Left-Hand Drive

The organ type accelerator pedal has welded at its base a rod which passes through the side panel of the scuttle, alongside the gearbox. A single lever is fixed to this rod by a clamp and bolt nut. From the end of this lever the movement of the pedal is transferred to a bell crank lever, mounted on the scuttle panel, by a rod having ball joint connectors at each end. A similar rod takes the movement forward to a second bell crank lever mounted on a spindle protruding from the exhaust manifold.

From this manifold bell crank a rod, with ball joint connectors at each end, transfers the pedal movement to a lever secured to the common throttle bar between the two carburetters.

As the disengagement of the overdrive is dependent, in some instances, upon throttle movement, an electrical switch is operated by a link rod extending upwards from the bell crank mounted on the scuttle.

Right-Hand Drive

As for the left-hand drive model the pedal mounting protrudes through the scuttle side panel where a single lever and a vertical rod take the movement upward. Here, instead of a bell crank lever, the movement utilizes

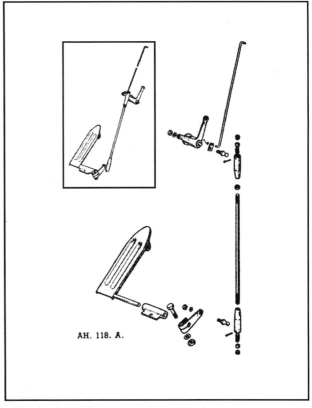

AH. 118. A.

Fig. 4. The left-hand drive accelerator pedal assembly in exploded form. The inset shows the components assembled.

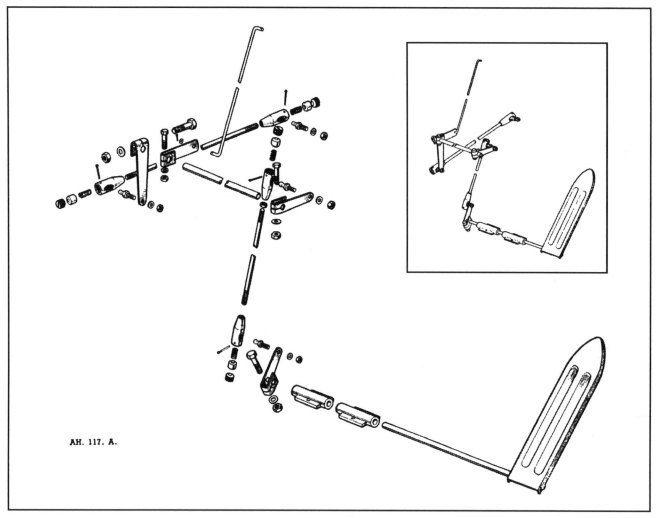

AH. 117. A.

Fig. 5. The right-hand drive accelerator pedal assembly and, inset, components assembled.

a single lever mounted on a transverse bar to take it across the engine compartment rear bulkhead to the left-hand side of the car, where two separate levers are mounted on the bar. One lever operates the overdrive switch via a vertical rod, the other lever is connected to the manifold bell crank lever by a double-ended ball joint connection. From this latter point the pedal move-ment is transferred to the common throttle bar of the carburetters by a ball joint connecting rod.

In both types of throttle operation a link rod from the choke control lever operates a cam which, in its turn, operates the throttle rod an amount sufficient to allow more fuel to flow through the throttle when the needle seat is lowered.

S.U. CARBURETTERS

General Description

Twin S.U. carburetters, inclined at an angle of 20 degrees to the horizontal, are fitted to the Austin-Healey "100". The S.U. Carburetter is of the auto-matically expanding choke type, in which the cross-sectional area of the main air passage adjacent to the fuel jet and the effective orifice of the jet, is variable. This variation takes place in accordance with the demand of the engine as determined by the degree of throttle opening, the engine speed and the load against which the engine is operating.

The distinguishing feature of this type of carburetter is that an approximately constant air velocity, and hence an approximately constant degree of depression, is at all times maintained in the region of the fuel jet. This velocity is such that the air flow demanded by the engine in order to develop its maximum power is not appreciably impeded, although good atomization of the fuel is assured under all conditions of load and speed.

The maintenance of a constant high air velocity across the jet, even under idling conditions, obviates the necessity for a separate idling jet. There is only one jet regulated by a needle (model Q.W.).

Construction

The main constructional features of the carburetter are shown in figs. 6 and 8.

These illustrations show the main body, butterfly throttle, automatically expanding choke and variable fuel-jet arrangement. They also indicate the means whereby the jet is lowered by a manual control to effect enrichment of the mixture for starting and warming up. The component parts of the float chamber are shown in fig. 7.

Turning to fig. 6, it will be seen that a butterfly throttle (3) mounted on a spindle is located close to the engine attachment flange, at one end of the main air passage and that an adjustable idling stop screw (3) fig. 8 is arranged to prevent complete closure of the throttle, thus regulating the flow of mixture from the carburetter under idling conditions with the accelerator released.

Towards the other end of the main passage is mounted the piston, its lower part constituting a shutter, restricting the cross-sectional area of the main air passage in the vicinity of the fuel jet as the piston falls. This component is enlarged at its upper end to form a piston of considerably greater diameter which moves axially within the bore of the suction chamber (2) fig. 6 and at the bottom of the piston is mounted the tapered needle which is retained by means of a setscrew (15).

The piston component is carried upon a control spindle which is slidably mounted within a bush fitted in the control boss forming the upper part of the suction chamber casting.

An extremely accurate fit is provided between the spindle and the bush in the suction chamber so that the enlarged portion of the piston is held out of contact with the bore of the suction chamber, within which, nevertheless, it operates with an extremely fine clearance. Similarly, the needle (1), fig. 10 is restrained from con tacting the bore of the jet (7) which it is seen to penetrate moving axially therein in correspondence with the rise and fall of the piston.

It will be appreciated that, as the piston rises, the air passage in the neighbourhood of the jet becomes enlarged and passes an additional quantity of air. Provided that the needle is of a suitably tapered form, its simultaneous withdrawal from the jet ensures the delivery to the engine of the required quantity of fuel corresponding to any given position of the piston and hence a given air flow.

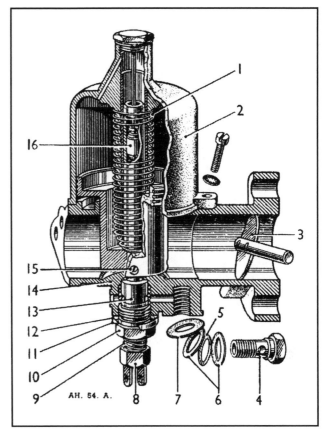

Fig. 6. Carburetter in sections.

1. *Piston spring.*
2. *Suction chamber.*
3. *Throttle butterfly valve.*
4. *Float chamber bolt.*
5. *Brass washer.*
6. *Small fibre washer.*
7. *Large fibre washer.*
8. *Jet adjusting nut.*
9. *Adjusting nut spring.*
10. *Jet holding screw.*
11. *Brass gland washer.*
12. *Gland.*
13. *Top half jet bearing.*
14. *Bridge.*
15. *Needle securing screw.*
16. *Hydraulic damper.*

The piston, under the influence of its own weight and assisted by the light compression spring (1) fig. 6, will tend to occupy its lowest position, a slight protuberance on its lower face contacting the bottom surface of the main air passage adjacent to the jet. The surface in this region is raised somewhat above the general level of the main bore of the carburetter and is referred to as the bridge (14) fig. 6.

Levitation of the piston is achieved by means of the induction depression, which takes effect within the suction chamber, and thus upon the enlarged portion of the piston through a drilling in the lower part of the piston which makes communication between this region and that lying between the piston and the throttle.

The annular space in the body immediately beneath the enlarged portion of the piston is completely vented to atmosphere by ducts.

It will be appreciated that, as the weight of the piston assembly is constant and the augmenting load of the spring approximately so, a substantially constant degree

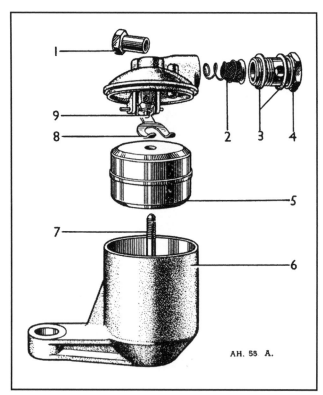

Fig. 7. Float chamber exploded.

1. *Cap nut.*
2. *Filter and spring.*
3. *Washers.*
4. *Fuel pipe union bolt.*
5. *Float.*
6. *Float chamber.*
7. *Float spindle.*
8. *Hinged fork.*
9. *Needle valve.*

drillings being provided therein to lead the fuel from the lower part of the float-chamber to the region surrounding the jet.

The buoyancy of the float, in conjunction with the form of the hinged fork is such that a fuel level is maintained approximately $\frac{1}{8}$ in. below the jet bridge. This can easily be observed after first detaching the suction chamber and suction piston then lowering the jet to its full rich position. The level can vary a further $\frac{1}{4}$ in. downward without any ill effects on the functioning of the carburetter.

The only parts of importance not so far mentioned are those associated with the jet.

Under idling conditions the piston is completely dropped, being then supported by a small spring-loaded protuberance provided on its lower surface, which is in contact with the bridge in the main bore of the carburetter body. The small gap thus formed between piston and bridge permits the flow of sufficient air to meet the idling demand of the engine without, however, creating enough depression on the induction side to levitate the piston.

The fuel discharge required from the jet is very small under these conditions, hence the diameter of that portion of the needle now obstructing the mouth of the jet is very nearly equal to the jet bore. Initial manufacture of the complete carburetter assembly to the required degree of accuracy to ensure perfect concentricity between the needle and the jet bore under these conditions is impracticable and an individual adjustment for this essential centralization is therefore provided.

It will be seen that the jet is not mounted directly in the main body, but is housed in the parts (9) and (10) (fig. 10) referred to as the jet bushes or jet bearings.

The upper jet bush is provided with a flange which forms a face seal against a recess in the body, while the lower one carries a similar flange contacting the upper surface of the hollow hexagon locking screw (5). The arrangement is such that tightening of the hollow hexagon locking screw will positively lock the jet and jet bushes in position. Some degree of lateral clearance is provided between the jet bushes and the bores formed in the main body and the locking screw. In this manner the assembly can be moved laterally until perfect concentricity of the jet and needle is achieved, the screw (5) being slackened for this purpose. This operation is referred to as "centreing the jet"; on completion of the operation the jet locking screw is finally tightened.

In addition to this concentricity adjustment, an axial adjustment of the jet is provided for the purpose of regulating the idling mixture strength.

As the needle tapers throughout its length, it will be clear that raising or lowering the jet within its bearings will alter the effective aperture of the jet orifice, and hence the rate of fuel discharge. To permit this adjustment the jet is slidably mounted within its bearings and provided with adequate sealing glands.

of depression will prevail within the suction chamber and consequently in the region between the piston and the throttle, for any given degree of lift of the piston between the extremities of its travel.

It will be clear that this floating condition of the piston will be stable for any given air-flow demand as imposed by the degree of throttle opening, the engine speed and the load; thus any tendency in the piston to fall momentarily will be accompanied by an increased restriction to air-flow in the space bounded by the lower side of the piston and the bridge and this will be accompanied by a corresponding increase in the depression between the piston and throttle, which is immediately communicated to the interior of the suction chamber, instantly counteracting the initial disturbance by raising the piston to an appropriate extent.

The float chamber, fig. 7, is of orthodox construction, comprising a needle valve (9) located with a separate seating which, in turn, is screwed into the float chamber lid and a float (5) the upward movement of which, in response to the rising fuel level, causes final closure of the needle upon its seating through the medium of the hinged fork (8).

It will be seen that the float chamber is a unit separate from the main body of the carburetter to which it is attached by means of the bolt (4) fig. 6, suitable

A compression spring (3) will be observed which, at its upper end, serves to compress the small sealing gland (2) and thus prevent any fuel leakage between the jet and the upper jet bearing.

In both locations a brass washer is interposed between the end of the spring and the sealing gland to take the spring thrust. A further sealing gland (11) together with a conical washer (4), is provided, to prevent fuel leakage between the jet screw (5) and the main body.

It will be seen from the diagram that the upward limit of slidable movement of the jet is moved upward toward the "weak" or running position.

The position of the nut (13) therefore determines the idling mixture ratio setting of the carburetter for normal running with the engine hot and it is prevented from unintentional rotation by means of the loading spring (12).

The cold running mixture control mechanism comprises a jet lever supported from the main body by a link member and attached by means of a clevis pin to the jet head (8). A tension spring is provided to assist in returning the jet moving mechanism to its normal running position.

Connection is made from the outer extremity of the jet lever to a control situated within reach of the driver.

Drillings in the float chamber attachment bolt (7), fig. 8, the main body of the carburetter, the jet (7) fig. 10, and slots in the upper jet bearing (9) serve to conduct fuel from the float chamber to the jet orifice.

One final feature of the carburetter may be noted from fig. 6. It will be seen that the spindle upon which the piston is mounted is hollow and that it surrounds a small stationary damper piston (16) suspended from the suction chamber cap by means of a rod. The hollow interior of the spindle contains a quantity of thin engine oil and the marked retarding effect upon the movement of the main piston assembly, occasioned by the resistance of this small piston, provides the momentary enrichment desirable when the throttle is abruptly opened. The damper piston is constructed to provide a one-way valve action which gives little resistance to the passage of oil during the downward movement of the main piston.

Adjustments and Fault Correction

The initial choice of the needle to suit the air/fuel requirements of the engine is determined, in the first instance, by careful observations of power and specific consumption on the test bed. It is not, therefore, a common requirement that the needle form be changed in practice, in cases where the marking on the needle coincides with the original maker's specification. Where any doubt exists as to the suitability of the needle form, this part may be withdrawn by removing the suction chamber and loosening the screw enabling the needle to be pulled out. Identification figures or letters will be found stamped either upon the outside of the needle shank, or upon its head.

When reinserting the needle, it is important that the shoulder, or junction between the parallel part of the needle, should coincide with the bottom of the piston rod into which it is inserted. After ascertaining the suitability of the needle form, tuning of the carburetter is generally confined to correct idling adjustment. This operation is carried out by means of the idling stop screw and the jet stop nut.

In the event of unsatisfactory behaviour of the engine and before proceeding to a detailed examination of the carburetter, it is advisable to carry out a check of the general condition of the engine.

Attention should, in particular, be directed towards the following :

(1) Incorrectly adjusted contact breaker gap, dirty or pitted contact points, or other ignition defects.
(2) Loss of compression of one or more cylinders.
(3) Incorrect plug gaps, oily or dirty plugs.
(4) Sticking valves, badly worn inlet valve guides.

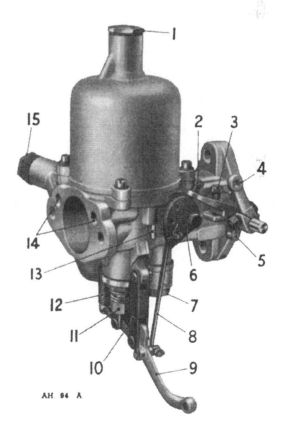

AH 94 A

Fig. 8. Carburetter control linkage.

1. Suction chamber cap.
2. Throttle adjusting screw.
3. Slow running screw.
4. Throttle lever.
5. Throttle stop.
6. Cam.
7. Float chamber bolt.
8. Connecting rod.
9. Jet lever.
10. Jet lever support link.
11. Jet adjusting nut.
12. Jet lever return spring.
13. Piston lifting pin.
14. Vents to atmosphere.
15. Fuel pipe union bolt.

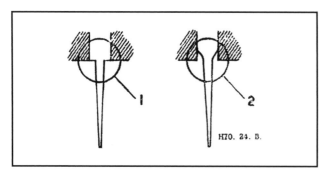

Fig. 9. Two forms of tapered needle— 1. Square shoulder. 2. Tapered shoulder. Although the needle with the tapered shoulder is not now manufactured, where stocks are supplied it may still be used.

(5) Defective fuel pump, or choked fuel filter.

(6) Leakage at joint between carburetter and engine flange, or between induction pipe flanges and cylinder head.

If no such defects are accountable for unsatisfactory engine performance it may be assumed that the carburation is at fault.

Sticking Piston

The symptoms here are either stalling and a refusal of the engine to run slowly or, alternatively, lack of power accompanied by excessive fuel consumption. This defect is easily detected.

It will be found that a projecting spring-loaded brass pin is provided beneath the suction chamber for testing the freedom of lift of the piston.

Should the piston be inclined to stick the air cleaner silencer should be removed to enable the operator to make a closer inspection through the intake throat.

The piston should rest, when the engine is not running, upon the bridge (14), fig. 6. When raised by means of the finger, or by the blade of a penknife, to its highest position against the appreciable resistance of the damper piston and then released, it should drop freely to strike the bridge sharply and distinctly.

Should the piston become prematurely arrested in its downward movement, or if it appears unduly reluctant to break away from its position of rest on the bridge when an attempt is made to raise it from this position, the jet should be lowered by means of the enrichment mechanism and the test repeated.

If the previous symptoms persist it can be assumed that the enlarged diameter of the piston is making contact with the bore of the suction chamber, or that the piston rod is not sliding freely within its bush. There is, furthermore, the possibility that the damper rod has become bent and is inducing friction between the damper piston and the bore of the main piston rod. This last possibility may be investigated by removal of the oil cap and damper assembly : if the removal of the damper allows

for downward movement of the piston the damper rod may then be straightened before reassembly. If, on the other hand, sticking has been eliminated by the act of dropping the jet, the indication is that contact and friction are taking place between the jet and the needle and that "centreing of the jet" is required.

Should dirt, or contact, between piston and suction chamber, or sticking of the piston rod in its bush be the cause of the trouble, remove the suction chamber, withdraw the piston and thoroughly clean both parts with petrol and a clean cloth.

Apply a few drops of light oil to the piston rod, preferably diluted with paraffin if any signs of rust or corrosion are noticed on the rod. Replace the piston in the suction chamber and test for rotational and sliding freedom.

Any direct local contact between these two parts, attributed to some indentation of the suction chamber, may be rectified by removing any high spots which may show up on the suction chamber bore, by careful use of a hand scraper. On no account should any attempt be made to enlarge generally the bore of the suction chamber, or to reduce the diameter of the enlarged part of the piston, as the maintenance of a limited clearance between these parts is absolutely essential to the proper functioning of the carburetter.

Eccentricity of the Jet and Needle

Re-centreing of the jet in relation to the needle will be necessary should the jet have become laterally displaced in service due to inadequate tightening of the locking screw (5), fig. 10, or any other cause. This operation will, of course, also be necessary if the jet and its associated parts have been removed for any reason. It may also be necessary after the replacement of a needle.

The procedure for re-centreing the jet is as follows : The jet stop nut (13) should first be screwed upwards to its fullest extent, and the jet head then raised to contact it so that the jet assumes its highest possible position. The locking screw (5) should now be loosened just sufficiently to release the jet and jet bush assembly (7), (9), (10), etc., and permit this to be moved laterally.

A moderate side loading applied to the lower protruding part of the bottom jet bush (10) will indicate whether or not the assembly has been sufficiently freed. The piston should now be raised and, maintaining the jet in its highest position, the piston should be allowed to drop. This will cause the needle to be driven fully into the jet mouth and thus bring about the required centralization. The locking screw should now be tightened and the jet returned to its former position. Should any indication of contact between the needle and the jet persist, which may sometimes occur due to further displacement of the assembly on finally tightening the locking screw, this must again be slackened off and the operation repeated until correct centralization has been achieved.

Flooding Float Chamber or Jet Mouth

Flooding may occur due to a punctured and petrol-laden float, or to dirt between the float-chamber needle valve and its seating. To remedy either defect, the float-chamber lid should be removed and the necessary cleaning, float replacement, or repair effected.

Leakage from Bottom of Jet

If persistent slow leakage is observed in the neighbourhood of the jet head, it is probable that the jet gland washer (2), fig. 10 and its lower counterpart together with the locking screw washer (11) require replacement.

The jet lever (9), fig. 8, should first be detached from the jet head, the locking screw (5) removed and the entire jet and jet bush assembly withdrawn.

On reassembly, great care should be taken to replace all parts in their correct positions as shown in the illustration. Re-centreing of the jet as previously described, will, of course, be necessary after this operation.

Water or Dirt in Carburetter

Should trouble due to this cause be suspected, the float chamber must be examined and cleaned out. If excessive water or foreign matter has been present, it is possible that the jet has become choked. Before removing the jet and its associated parts for cleaning, the following expedient may be attempted.

The jet should be dropped to its full extent, the suction chamber and piston removed and the suction chamber alone replaced. The main air inlet should then be obstructed and the engine rapidly, but briefly, turned over by hand, or by the starter. This operation subjects the jet to a high degree of suction which will probably result in any foreign matter being drawn out, both from the jet and from adjacent fuel passages. Should this operation fail, however, the jet and associated parts must be removed for cleaning.

Fuel Supply Failure at Float Chamber

If the engine is found to stop under idling or light running conditions, notwithstanding the fact that a good supply of fuel is present at the float chamber inlet union, it is possible that the needle has become stuck to its seating.

This possibility arises in the rare cases where some gummy substance is present in the fuel system. The most probable substance of this nature is the polymerized gum which sometimes results from the protracted storage of fuel in the tank.

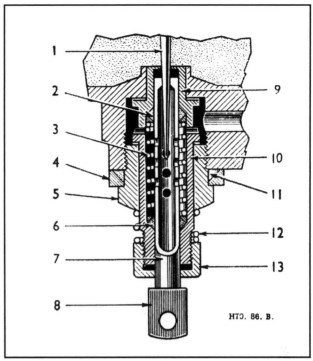

Fig. 10.   Sectional view of Jet.

1. *Tapered needle.*	7. *Jet.*
2. *Jet gland.*	8. *Jet head.*
3. *Gland spring.*	9. *Upper jet bush.*
4. *Locking screw gland washer.*	10. *Bottom jet bush.*
5. *Locking screw.*	11. *Locking screw gland.*
6. *Lower jet gland.*	12. *Adjusting nut spring.*

13. *Jet stop adjusting nut.*

After removal of the float chamber lid and float lever, the needle may be withdrawn and its point thoroughly wiped with a rag dipped in alcohol. Similar treatment should also be applied to the needle seating, which can be conveniently cleaned by means of a matchstick and cloth dipped in alcohol. Persistent trouble of this nature can only be cured by completely stripping down the fuel system and thoroughly cleansing each component and pipe line.

If the engine is found to suffer from a serious lack of power, which becomes evident at higher speeds and loads, this is probably due to an inadequately sustained fuel supply when the fuel pump should be investigated for inadequate delivery and its filter cleaned and inspected.

Sticking Jet

Should the jet and its operating mechanism become unduly resistant to the action of lowering and raising by means of the enrichment mechanism, the jet should be lowered to its full extent and the lower part thus exposed should be smeared with petroleum jelly or similar lubricant. Oil should also be applied to the various linkage pins in the mechanism and the jet raised and lowered several times in order to promote the passage of the lubricant upwards between the jet and its "entourage".

Throttle and Mixture Control Interconnection

A direct connection is provided between the jet movement and the throttle opening. Such an interconnection ensures that the engine will continue to run when the mixture is enriched by lowering the jet, without the additional necessity of maintaining a greater throttle opening than is normally provided by the setting of the slow running screw.

The mechanism involved in this interconnection is shown in fig. 8. It will be seen that a connecting rod conveys movement from the jet lever to a cam pivoted on the side of the main body casting.

Movement of the jet lever in the direction of enrichment is thus accompanied by an upward movement of the cam which, in turn, abuts against the adjustable screw and this opens the throttle to a greater degree than the normal slow-running setting controlled by the slow-running stop-screw. The throttle adjusting screw should be so adjusted that it is just out of contact with the cam when the jet has been raised to its normal running position, and the throttle is shut back to its normal idling condition as determined by the slow-running screw.

Tuning Twin Carburetters

To make a thorough job of adjusting twin S.U. carburetters it is first advisable to check all engine details which affect performance, such as tappet clearances, plug gaps and distributor gap. The carburetters should then be checked over in accordance with the instructions given in section "Adjustments and Fault Correction", making sure that the pistons are perfectly free and that the jets are correctly centred.

Now slacken the clamping bolts on the universally jointed connections between the throttle spindles so that the throttles can be set independently and disconnect the mixture control linkage by removing one of the fork swivel pins. While the suction chambers are off, see that the needles are located in the same position in all the pistons and that the jets are the same distance below the bridges of the carburetters when they are pushed hard against their adjusting nuts.

Unscrew the throttle-adjusting screws and screw these back until they will just hold a piece of thin paper inserted between the adjusting screw and the stop lug, then screw them in one complete turn. The engine may now be started. When it is thoroughly warmed up the speed may be adjusted by turning the throttle-adjusting screws equal amounts in either direction, depending on whether a higher or lower speed is required. To check for exact synchronization of the throttle openings it is best to listen to the intake. This is most easily done by holding one end of a piece of rubber tubing against the ear and holding the other end near the intake of each of the carburetters in turn. If the hiss on one of them is louder than the other, unscrew its throttle-adjusting screw until the intensity of the hiss is equal.

When it is obvious that this is satisfactory, the mixture should be adjusted by screwing the jet adjusting nuts up or down to exactly the same extent, pushing the jets hard against the nuts until satisfactory running is obtained.

As these are adjusted the engine will probably run faster and it may therefore be necessary to unscrew the throttle-adjusting screws a little, each by the same amount, in order to reduce speed.

When the mixture is correct on both carburetters, lifting the piston of one of them with a penknife blade should make the engine beat become irregular from excessive weakness. If lifting the piston on one carburetter stops the engine and lifting that of the other does not, this indicates the mixture on the first carburetter is set weaker than that on the second and the first one should therefore be enriched by unscrewing the jet-adjusting nut. Once the mixture is correct from both carburetters the exhaust beat should be regular and even. If it is irregular, with a spashy type of misfire and a colourless exhaust, the mixture is too weak. If there is a regular or rhythmical type of misfire in the exhaust beat, together with a blackish exhaust, then the mixture is too rich.

Before re-connecting the mixture control linkage, make sure that the jets are hard up against the adjusting nuts and, if necessary, adjust the length of the linkage so that the clevis pins may be inserted freely while the jets are in this position. The throttle spindle interconnection clamping bolts may now be tightened.

FUEL PUMP

Description

The S.U. electric fuel pump is the "L" type high-pressure pump which is capable of a continuous delivery of eight gallons of petrol per hour through a suction lift of four feet.

Mounted beneath the car on the rear bulkhead of the cockpit the pump has its filter gauze situated at the bottom of the body whilst the inlet and outlet unions are at the top. The pump consists of three main assemblies—the body, the magnet assembly and the contact breaker.

The body is composed of two aluminium die castings, (17) and (19), of fig. 11, with a fabric joint washer between them. The outlet union (9) tightens down on to the delivery cage (13) which is clamped

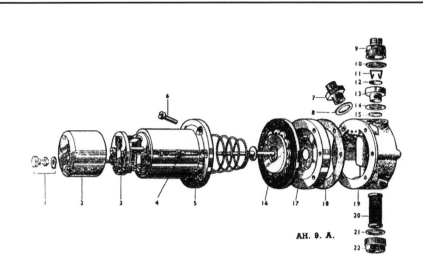

Fig. 11. Fuel pump in exploded form.

1. Moulded terminal knob, locknut and washer.
2. End cover cap.
3. Contact breaker assembly.
4. Earthing stud.
5. Magnet housing.
6. Body fixing screw.
7. Inlet union.
8. Fibre washer.
9. Outlet union.
10. Fibre washer.
11. Valve disc retainer clip.
12. Delivery valve disc.
13. Valve cage.
14. Fibre washer.
15. Suction valve disc.
16. Diaphragm assembly.
17. Body plate.
18. Joint washer.
19. Body casting.
20. Filter washer.
21. Fibre washer.
22. Filter plug.

AH. 9. A.

between two fibre washers. In the top of the cage is the delivery valve, a thin brass disc (12), held in position by a spring clip (11); the suction valve (14) being a similar disc resting on a seating machined in the pump body. Holes connect the space between the valves to the pumping chamber which is a shallow depression on the forward face of the smaller body casting. This space is closed by a diaphragm assembly (16) which is clamped at the outside by the magnet housing (5) and the pump body. A bronze rod is screwed through the centre of the armature to which the diaphragm is attached and passes through the magnet core to the contact breaker which is located at the far end. A spring is interposed between the armature and the end plate of the coil.

The magnet consists of a cast-iron pot having an iron core on which is wound a coil of copper wire that energizes the magnet. Between the magnet housing and the armature are fitted eleven spherical-edged brass rollers. These rollers locate the armature centrally within the magnet in a longitudinal direction. The contact breaker assembly (3) fig. 11, consists of a small bakelite moulding carrying two rockers which are both hinged to the moulding at one end and are connected together at the top by two small springs arranged to give a throw-over action. A trunnion is fitted into the centre of the inner rocker and the bronze rod connected to the armature is screwed into this. The outer rocker is fitted with a tungsten point which makes contact with a further tungsten point on a spring blade. This spring blade is connected to one end of the coil whilst the other end is connected to a terminal.

There is a short length of flexible wire connecting the outer rocker to one of the screws holding the bakelite moulding to the magnet housing, thus ensuring a good earth.

Operation

When the pump is at rest the outer rocker lies in the outer position and the tungsten points are in contact. The current passes from the terminal, through the coil, back to the blade, through the points and to earth, thus energizing the magnet and attracting the armature. This comes forward, bringing the diaphragm with it and sucking petrol through the suction valve into the pumping chamber. When the armature has advanced nearly to the end of its stroke, the throw-over mechanism operates and the outer rocker flies back, separating the points and breaking the circuit. The large coil spring then pushes the armature and the diaphragm back, forcing petrol through the delivery valve at a rate determined by the requirements of the engine. As soon as the armature gets near the end of this stroke the throw-over mechanism again operates, the points make contact once more and the cycle of operations is repeated.

Maintenance

Should fuel trouble arise and in consequence the pump be suspected, the delivery pipe should be disconnected at the pump union to ascertain that the pump is working.

If the failure is attributed to the pump, disconnect the lead from the terminal and strike it against the pump body to verify that it sparks, thus proving whether or not there is current available in the wire. With current and yet no working of the pump, the bakelite cover should be removed and the terminal touched with the lead. If the pump does not operate and the points are in contact and yet a spark cannot be struck off the terminal, it is probable that there is some dirt on the points.

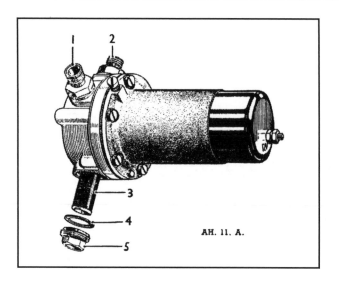

Fig. 12. Fuel pump.
1. Inlet union. 2. Outlet union. 3. Filter gauge. 4. Fibre washer. 5. Filter cap nut.

The points may be cleaned by inserting a piece of card between them, pinching them together and sliding the card backwards and forwards. If, when the wire is connected to the terminal, the points fail to break, the possible cause is an obstruction of the suction pipe which may be cleared by blowing down the pipe with a tyre pump.

Another possible cause of the points failing to open is a hardened diaphragm or possibly foreign matter in the roller assembly which supports the diaphragm. In the latter case the diaphragm should be removed and the whole assembly cleaned and reassembled.

Should the pump become noisy, look for an air leak on the suction side. The simplest way to check for this is to disconnect the petrol pipe from the carburetters and allow the pump to force petrol into a pint can. Then submerge the end of the pipe in the petrol. If bubbles come through the pipe there must be an air leak which must be found and cured.

Should the pump keep beating without delivery of fuel, it is possible that a piece of dirt is lodged under one of the valves.

These can be removed by unscrewing the top union and lifting the cage out. When replacing the cage ensure that the thin hard red fibre washer is below it and the thick orange-coloured washer above. A choked filter or an obstruction on the suction side will make the pump very hot eventually causing failure.

Occasionally the gap area of the contact points should be checked. The spring blade rests against a small projection on the bakelite moulding and it should be set so that when the points are in contact it is deflected back from the moulding. The width of the gap at the points is approximately .030 inch.

AIR CLEANER—SILENCER

The Burgess air cleaner-silencers, two in number, fitted to the S.U. Carburetters are of the oil-wetted type and apart from regular cleaning require little or no attention.

To withdraw the element from an air cleaner, first extract the central setpin, with its shakeproof washer, from the top cover plate. Then lift off the plate and pull out the gauze element. The element, when cleaning, should be swilled in petrol, drained, then immersed in engine oil and again drained before being refitted.

To remove the cleaner-silencer as a whole its backplate must be released from the carburetter flange concerned where two setpins secure the cleaner-silencer in place.

Where the front air cleaner is concerned the valve rocker cover breather hose must be pulled clear of its connection at the cleaner-silencer backplate.

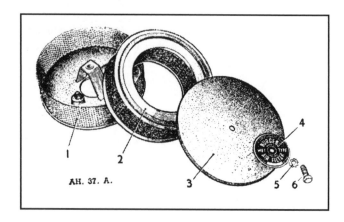

Fig. 13. The air cleaner in exploded form.
1. Outer case. 2. Filter element. 3. Top cover. 4. Name plate. 5. Shakeproof washer. 6. Setpin.

SERVICE DIAGNOSIS GUIDE

Symptom	No.	Possible Cause
(a) High Fuel Consumption	1 2 3 4 5 6 7	In Section D, pages 27/28:— Check 2 and 3 in (a) Check 1, 2 and 6 in (b) Check 1 in (d) Check 3 in (g) Over-rich mixture Dirty air cleaner (page C/12) Excessive vibration on engine mountings Flooding carburetter, see (b) Incorrect jet setting (page C/8) Leaks in fuel supply line Strangler flap partially 'on'
(b) Flooding Carburetter	1 2 3 4 5 6	Needle valve sticking Needle valve not seating properly Damaged float chamber gasket Loose float chamber retaining bolts Punctured float Incorrect pump setting
(c) Weak Mixture	1 2 3 4 5 6	Under-size jets Slow-running adjustment incorrect Leaking or damaged carburetter joint washer Warped or damaged carburetter flange Leaking or damaged manifold joint washer Loose manifold nuts
(d) Fuel Gauge Reading Incorrect	1 2 3 4	Faulty fuel gauge Imperfect contact at tank unit or gauge connections Broken lead to tank unit, or lead disconnected Tank unit lead shorting

AUSTIN HEALEY '100' ENGINE

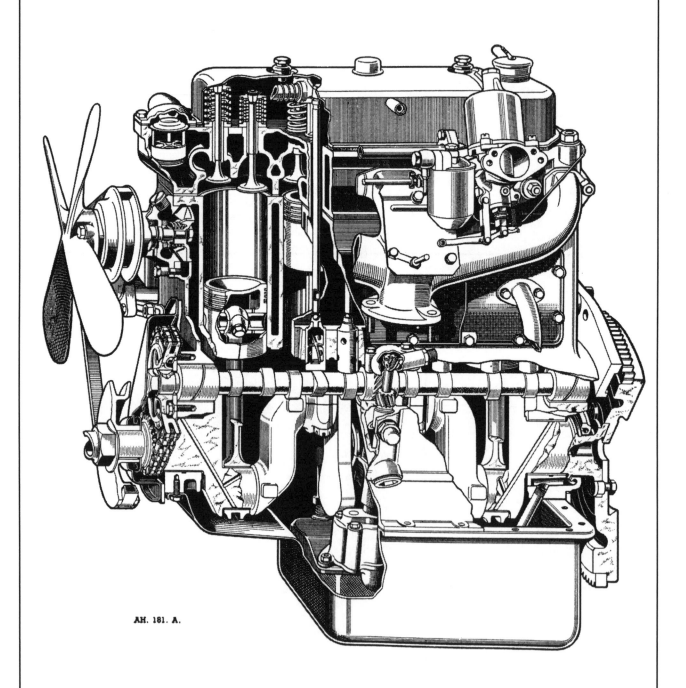

AH. 181. A.

Fig. 1. A cut-away view of the Austin Healey '100' engine.

ENGINE

TECHNICAL DATA

General

Number of Cylinders	4
Bore	3.4375 in. (87.3 mm.)
Stroke	4.375 in. (111.1 mm.)
Capacity	162.42 cu. in. (2,660 cc.)
Brake Horse Power	90 (91.248 cv.) at 4,000 r.p.m.
Torque	150 lbs./ft. (20.74 kgm.) at 2,000 r.p.m.
B.M.E.P. Max	139 lbs./sq. in. (9.773 kg./cm².) at 2,000 r.p.m.
R.A.C. Rating	18.91 H.P.
Firing order	1, 3, 4, 2
Valve Position	Overhead Push Rod Operation
Compression ratio	7.5 to 1

Torque Wrench Settings

Cylinder Head Nuts	65–70 lbs./ft. (8.987–9.678 kgm.)
Main Bearing Stud Nuts	Castellated-tighten to Nr. Slot
Connecting rod big end nuts	Castellated-tighten to Nr. Slot
Connecting rod small end clamping screw	
Flywheel crankshaft bolts	35–40 lbs./ft. (4.839–5.530 kgm.)
Front Cover Setpins	
Front Mounting plate setpins	
Rear Mounting plate setpins	
Engine support bracket bolts to front plate	

Cylinder Liners

Diameter of bore to receive liner	3.537–3.537½-in. (89.839–89.852 mm.)
Outside diameter of liner	3.539–3.539¾-in. (89.890–89.903 mm.)
Finished bore of liner	3.437–3.438½-in. (87.299–87.347 mm.)

Pistons

Material	Aluminium Alloy
Piston clearance at skirt. (Measured at right angles to the gudgeon pin)	.0012–.0018-in. (.030–.045 mm.)
Width of ring groove :—	
Compression	.0957–.0967-in. (2.430–2.456 mm.)
Oil Control	.158–.159-in. (4.013–4.038 mm.)
Compression rings :—	
1st groove	Plain
2nd and 3rd grooves	Taper
Oil Control :—	
4th groove	Slotted Scraper
Ring Width :—	
Compression	.093–.094-in. (2.362–2.387 mm.)
Oil Control	.1552–.1562-in. (3.942–3.967 mm.)
Clearance in groove :—	
Compression	.0017–.0037-in. (.043–.093 mm.)
Oil Control	.0018–.0038-in. (.045–.096 mm.)
Ring Gap	.011–.015-in. (.279–.381 mm.)

Piston and rings are available in .010-in. (.2121 mm.), .020-in. (.508 mm.), .030-in. (.762 mm.) and .040-in. (1.016 mm.).

The diameter of the piston is measured at right angles to the gudgeon pin at the bottom of the skirt.

Gudgeon Pin

Type	Floating in Piston
Fit	Thumb fit at 70°F. (21.1°C.)
Length	$3\frac{1}{16}$-in. (7.778 cm.)
Diameter	.8748–.8750-in. (22.219–22.225 mm.)

Crankshaft

Material	Steel Forging
Diameter of main journals	2.479–2.479½-in. (62.966–62.979 mm.)
Diameter of crankpins	2.000–2.0005-in. (50.800–50.812 mm.)
Crankshaft end float	.002–.003-in. (.0508–.0762 mm.)
Main journal clearance	.001–.0025-in. (.025–.0762 mm.)

Main journal regrind sizes :—

1st undersize—.010-in. (.254 mm.)	2.4690–2.469½-in. (62.712–62.725 mm.)
2nd undersize—.020-in. (.508 mm.)	2.4590–2.459½-in. (62.458–62.217 mm.)
3rd undersize—.030-in. (.762 mm.)	2.4490–2.449½-in. (62.204–62.217 mm.)
4th undersize—.040-in. (1.016 mm.)	2.4390–2.439½-in. (61.950–61.967 mm.)

Crankpin regrind sizes :—

1st undersize—.010-in. (.254 mm.)	1.9900–1.9905-in. (50.546–50.558 mm.)
2nd undersize—.020-in. (.508 mm.)	1.9800–1.9805-in. (50.292–50.304 mm.)
3rd undersize—.030-in. (.762 mm.)	1.9700–1.9705-in. (50.038–50.050 mm.)
4th undersize—.040-in. (1.016 mm.)	1.9600–1.9605-in. (49.784–49.796 mm.)
End float taken on	Washer at centre bearing

Connecting Rods

Material	Steel Forging
Length (centre to centre)	8.183–8.187½-in. (20.7–20.8 cm.)
Big end housing diameter	2.114–2.114½-in. (53.595–53.608 mm.)
Side clearance	.008–.012-in. (.2032–.3048 mm.)
Big-end bearing clearance	.0005–.002-in. (.0127–.0508 mm.)

Main Bearings

Type	Thin wall, steel-backed white metal
Number	3
Length—Front	1.745–1.755-in. (44.323–44.577 mm.)
Centre	1.745–1.755-in. (44.323–44.577 mm.)
Rear	1.990–2.000-in. (50.546–50.800 mm.)
Inside Diameter	2.4805–2.4815-in. (63.004–63.030 mm.)
Thickness	.072–.072¼-in. (1.828–1.835 mm.)
Housing Diameter	2.625–2.6255-in. (66.675–66.687 mm.)

Big End Bearing

Type	Thin wall white metal lined
Overall Length	1.226–1.236-in. (31.240–31.394 mm.)
Inside Diameter	2.001–2.002-in. (50.825–50.850 mm.)
Thickness	.056¼–.056½-in. (1.428–1.435 mm.)

Camshaft

Material	Steel Forging
Journal Diameters :—	
Front	1.788¾–1.789¼-in. (45.434–45.446 mm.)
Centre	1.748¾–1.749¼-in. (45.418–45.430 mm.)
Rear	1.622¾–1.623¼-in. (41.217–41.230 mm.)
Bearing Clearance (all three)	.001–.002-in. (.0254–.0508 mm.)
End Float	.003–.006-in. (.0762–.1524 mm.)

Camshaft—continued
　　End Thrust Taken on ... ... Locating Plate
　　Drive ... ... Duplex Roller Chain
　　Chain Pitch ... ... .375-in. (9.525 mm.)
　　Number of Pitches 62

Camshaft Bearings
　　Type ... ... Thin wall white metal lined
　　Number 3
　　Front Bearing :—
　　　　Housing diameter ... ... 1.915–1.916-in. (48.641±48.666 mm.)
　　　　Outside diameter (before fitting) ... ... 1.920±.0005-in. (48.768±.0127 mm.)
　　　　Inside diameter (reamed in position) ... ... 1.790¼–1.790¾-in. (45.472–45.485 mm.)
　　　　Length ... ... 1 11/16-in. (42.862 mm.)
　　Centre Bearing :—
　　　　Housing diameter ... ... 1.875–1.876-in. (47.625–47.650 mm.)
　　　　Outside diameter (before fitting) ... ... 1.880-in. (47.752 mm.)
　　　　Inside diameter (reamed in position) ... ... 1.750¼–1.750¾-in. (44.456–44.469 mm.)
　　　　Length ... ... 1¾-in. (44.45 mm.)
　　Rear Bearing :—
　　　　Housing diameter ... ... 1.749–1.750-in. (44.424–44.500 mm.)
　　　　Outside diameter (before fitting) ... ... 1.754-in. (44.451 mm.)
　　　　Inside diameter (reamed in position) ... ... 1.624¼–1.624¾-in. (41.255–41.268 mm.)
　　　　Length ... ... 1⅝-in. (41.275 mm.)

Valves
　　Material—Inlet ... ... Silicon Chrome Steel
　　　　　　Exhaust ... ... XB Steel

Timing
　　Inlet Opens ... ... 5° B.T.D.C.
　　Inlet Closes ... ... 45° A.B.D.C.
　　Exhaust Opens ... ... 40° B.B.D.C.
　　Exhaust Closes ... ... 10° A.T.D.C.
　　Valve clearance for Setting Timing ... ... .021-in. (.534 mm.)
　　Head diameter :
　　　　Inlet ... ... 1.725–1.730-in. (43.700–43.943 mm.)
　　　　Exhaust ... ... 1.415–1.420-in. (35.941–36.068 mm.)
　　Seat Angle :
　　　　Inlet and Exhaust ... ... 45°
　　Seat Width :
　　　　Inlet (approx.) ... ... 1/16-in. (1.5875 mm.)
　　　　Exhaust (approx.) ... ... ⅛-in. (3.175 mm.)
　　Stem Clearance :
　　　　Inlet ... ... .0015½–.0025½-in. (.0393–.0647 mm.)
　　　　Exhaust ... ... .001–.002-in. (.0254–.0508 mm.)
　　Stem Diameter :
　　　　Inlet ... ... .341¾–.342¼-in. (8.680–8.693 mm.)
　　　　Exhaust ... ... .341¾–.342¼-in. (8.680–8.693 mm.)
　　Lift ... ... .390-in. (9.906 mm.)
　　Length (overall) :
　　　　Inlet ... ... 4 25/32-in. (12.144 cm.)
　　　　Exhaust ... ... 4 21/32-in. (11.826 cm.)
　　Working Clearance ... ... 012-in. (.3048 mm.)

Valve Guides

Length :

Inlet ... ... $2\frac{2}{14}$-in. (5.437 cm.)

Exhaust ... ... $2\frac{11}{14}$-in. (7.183 cm.)

Outside Diameter :

Inlet and Exhaust ... ... .5635–.5640-in. (14.312–14.325 mm.)

Inside Diameter :

Inlet ... ... .3438–.3448-in. (8.732–8.757 mm.)

Exhaust ... ... .3433–.3438-in. (8.719–8.732 mm.)

Height above valve spring Seat ... ... $\frac{11}{16}$-in. (17.462 mm.)

Valve Springs

Outer :

Free Length... ... ... $2\frac{11}{64}$-in. (5.516 cm.)

Fitted Length and Load ... ... $1\frac{11}{16}$-in. at 65 ± 2 lbs. (4.325 cm. at $29.484\pm.907$ kgm.)

Number of Working Coils ... ... $4\frac{1}{2}$

Diameter of Wire ... ... .176-in. (4.470 mm.)

Core Diameter ... ... 1.125–1.140-in. (28.675–28.956 mm.)

Inner :

Free Length... ... ... $1\frac{61}{64}$-in. (4.960 cm.)

Fitted Length and Load ... ... $1\frac{1}{2}$-in. at $22\frac{1}{2}\pm1$ lb. (3.810 cm. at $10.205\pm.454$ kgm.)

Number of Working Coils ... ... $6\frac{1}{2}$

Diameter of Wire ... ... .110-in. (2.794 mm.)

Core Diameter ... ... .750–.765-in. (19.050–19.431 mm.)

Tappets

Type ... ... Barrel, dome base

Diameter ... ... .998$\frac{3}{4}$–.999$\frac{1}{4}$-in. (25.368–25.380 mm.)

Length (overall) ... ... 3.026$\frac{1}{4}$–3.036$\frac{1}{4}$-in. (76.866–77.120 mm.)

Rocker Mechanism

Push Rods :

Overall Length ... ... 10.130–10.155-in. (25.730–25.793 cm.)

Stem Diameter ... ... $\frac{3}{8}$-in. (9.525 mm.)

Rocker Shaft :

Length ... ... 16$\frac{1}{8}$-in. (40.957 cm.)

Outside Diameter ... ... .810–.811-in. (20.574–20.599 mm.)

Rocker arm bush :

Type ... ... Thin wall white metal lined

Outside Diameter (before fitting) ... ... .913–.914-in. (23.190–23.215 mm.)

Inside Diameter (reamed in position) ... ... .8115–.8125-in. (20.612–20.637 mm.)

Thickness before reaming ... ... .0525–.0540-in. (1.333–1.371 mm.)

Clearance ... ... .0005–.0025-in. (.0127–.0635 mm.)

Rocker Arm :

Bore ... ... .909–.910-in. (23.088–23.144 mm.)

Ratio 1.42

Flywheel

Material ... ... High Tensile Cast Iron

Diameter ... ... 12$\frac{13}{16}$-in. (32.543 cm.)

Starter ring :

Number of Teeth 106

Diameter over Teeth ... ... 13.292–13.297-in. (33.761–33.774 cm.)

Lubrication

Type of System	Forced Feed
Type of Pump	Straight Gear
Oil Pressure	... 50–55 lbs. per sq. in. (3.515–3.867 kg./cm².)
External Filter:	
Type	... Full Flow
Make	Tecalemit or Purolator
Capacity	1¼ pints (.71 litre)
Sump Capacity	11¾ pints (6.68 litres)

LUBRICATION

LUBRICATION is of extremely vital importance to ensure the reliability and long life of the moving parts. Great care has been taken to select oils which will give the best results under all operating conditions. It is therefore imperative that the correct grades of oil be used and that they should be applied in accordance with a definite schedule. The chart given in Section U should be regularly referred to for details of mileage application, grade and quantity of lubricant required.

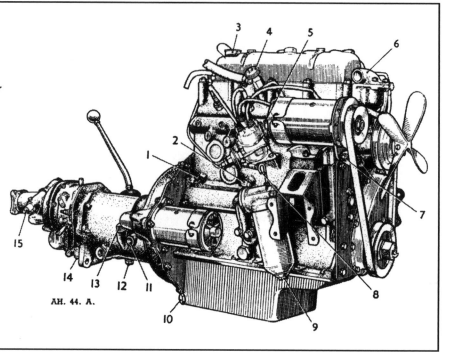

Fig. 2. Right-hand side view of Power Unit.

1. *Oil pressure gauge union.*
2. *Cylinder block drain tap.*
3. *Oil reservoir filler cap.*
4. *Heater valve.*
5. *Dynamo lubricator.*
6. *Thermostat cover.*
7. *Fan belt adjusting nuts.*
8. *Dipstick.*
9. *Oil filter securing bolt.*
10. *Oil reservoir drain plug.*
11. *Clutch operating shaft.*
12. *Gearbox drain plug.*
13. *Gearbox filler plug.*
14. *Overdrive drain plug.*
15. *Speedometer drive connections.*

AH. 44. A.

Description

There is full pressure lubrication throughout the unit.

The gear type pump draws oil from the sump through a gauze oil filter and delivers it to all bearings and the camshaft chain.

The sump capacity is 11¾ pints, but an external full flow oil filter is fitted, which must be charged separately, and takes an additional 1¼ pints.

The oil filler is in the valve cover on top of the cylinder head, and the oil level is checked by a dipstick which is on the right-hand side of the engine.

Draining the Sump

The engine and oil filter should be completely drained and fresh oil put in at least every 3,000 miles to provide the best possible running conditions.

There is a drain plug in the base of the sump. On new or reconditioned engines draining should be done after the first 500 miles running and again after the next 3,000 miles. After this period no further attention need be given to the filter, except the renewal of the element as described, when necessary.

Drain when the engine is warm and under no circumstances should petrol or paraffin be poured through the oil filter to clean the engine.

Refilling

When refilling, do not pour the oil in too fast, otherwise it may overflow through the breather at the front end of the valve cover. Check periodically that this breather is not choked up. Failure to keep this clear may result in condensation on the valve gear.

Test the level of the oil with the dipstick, wiping the stick clean before taking the reading. This should only be done when the vehicle is on level ground and not immediately after the engine has been run, or a false reading may be given.

Circulation

The oil circulation is clearly shown on page S/3. Starting at the gauze filter and pick-up in the sump, oil is drawn into the pump, from which it is fed to the full flow oil filter and thence to the main oil gallery. This runs the length of the engine on the right-hand side, from which the main oil delivery is made. A spring-loaded oil release valve, located between the pump and filter and accessible from the exterior of the crankcase, is provided, the overflow from which is returned to the sump filter. From the main oil gallery, oil is fed to the big ends, main bearings and the three camshaft bearings.

From one camshaft bearing, oil at reduced pressure is taken through drilled passages in the cylinder block and cylinder head to an oil-feed collar on the valve rocker shaft, and thence to the drilled shaft itself. Therefore the shaft is under pressure, surplus oil after circulation returning from the rocker gear via the push rod holes to the sump.

At the front end of the front camshaft bearing there are two oil bleed holes which feed oil to the camshaft gear and thence to the timing chain. See Fig. 4. Separate lubrication for the cylinder bores is effected by a small jet hole in the top half of each connecting rod big end bearing.

Oil Pressure Gauge

The oil pressure gauge gives an indication whether the oiling system is working properly. The normal oil pressure during ordinary running should be not less than 50 to 55 lbs. per square inch, with a proportionate lower pressure when idling, and will keep constant as long as the filter element remains clear and is not choked. As the filter gradually becomes choked, the oil pressure progressively becomes less. A drop to between 30–35 lbs. per square inch is an indication that the element is being by-passed and that it should be renewed to restore the oil pressure to normal.

The gauge should be observed when the engine is first started up after refilling the sump to check that the oil is circulating and that the pressure is correct. It should also be kept under observation frequently during normal running. Should the gauge fail to register a normal pressure, it may be due to lack of oil in the crankcase. If oil is present and the gauge still fails to register, stop

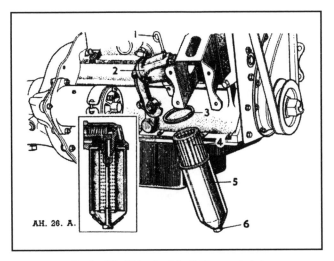

Fig. 3. The "Tecalemit" oil filter exploded.

1. *Dipstick.* 2. *Filter head.* 3. *Sealing ring.* 4. *Filter element.* 5. *Filter body.* 6. *Securing bolt. Inset shows alternative Micronic "Purolator" Filter.*

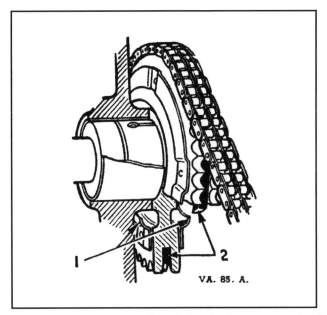

Fig. 4. Oil catchers 1 feed a flow of oil to the chain on each side of the camshaft and a synthetic rubber tensioner ring 2 ensures quiet running and takes up slack in the chain.

the engine immediately and check for a broken pipe or other cause. Test the gauge by a replacement, clamped direct to the instrument panel.

Check for Loss of Pressure

First, check the sump oil level by means of the dipstick. If the level is well up, check the oil gauge pipe from crankcase to instrument panel for fracture or leak. If the pipe is in order, remove the sump and examine the gauze filter. This may be choked; also remove release valve and inspect for foreign matter.

If these tests fail to indicate the cause of the loss of pressure or oil circulation, the crankshaft and other bearings will have to be closely examined and stripped down if necessary.

"Tecalemit" Filter

The external filter is of a full flow type, thus ensuring that all oil in the lubrication circuit passes through the filter before reaching the bearings.

The element of the filter is of a star formation in which a special quality felt, selected for its filtering properties, is used.

Oil is passed to the filter from the pump at a pressure controlled at 50/55 lbs. per square inch by the engine oil release valve. This pressure will, of course, be somewhat higher until the oil reaches a working temperature. Some pressure is lost in passing the oil through the filter element; this will only be a pound or two per square inch with a new element, but will increase as the element becomes progressively contaminated by foreign matter removed from the oil.

Should the filter become completely choked due to neglect, a balance valve is provided to ensure that oil will still reach the bearings. This valve, set to open at a pressure difference of 15/20 lbs. per square inch, is non-adjustable and is located in the filter head casting. When the valve is opened, unfiltered oil can by-pass the filter element and reach the bearings.

To renew the filter element proceed as follows:—

(1) Stop the engine, extract the centre fixing bolt, remove the container and drain.

(2) Withdraw the contaminated element and carefully cleanse the container of all foreign matter that has been trapped.

(3) After ensuring that no fibres from the cleansing operation **have been left in the container**, put in a new element, hold centre bolt firm, prime the filter, and refit to head casting, tightening the centre fixing bolt sufficiently to make an oil-tight joint and then top up the engine with oil.

It is highly recommended that the filter container should not be disturbed other than for the fitting of a new element; to do so invites the hazard of added contamination from accumulated dirt on the outside of the filter entering the container and thus being carried into the bearings on restarting the engine.

Micronic Filter

The micronic "Purolator" filter is sometimes fitted as an alternative to the "Tecalemit" filter, although in function they are identical.

In the same way as the "Tecalemit" filter, oil is passed from the pump at a pressure controlled by the engine release valve and it also has a balance valve fitted in the filter head.

The element should be handled carefully and changed every 9,000 miles or as soon as the oil begins to discolour; whichever is the sooner.

To remove the element, clean the exterior of the filter assembly and then unscrew the centre bolt, see fig. 3, withdraw the container and element.

Remove the filter element and thoroughly clean the interior of the container. Ensure that the rubber seal is in good condition and in position within the filter head. Place the new element on the lower element guide and offer up the container complete with element and fill with new oil to the filter head so that the former seats squarely on the seal and the latter is located on the upper element guide.

Screw the centre bolt back into the centre tube firmly enough to ensure that there will be no leakage past the seals.

Top up with oil and then run the engine for a few minutes and inspect for leakage.

To Remove the Oil Sump

First, drain off the oil by taking out the drain plug;

the oil capacity is approximately 11¾ pints, excluding full flow filter.

The sump is secured in position with 22 screws. Support the sump while removing these screws and then carefully lower clear of the oil pump gauze strainer and pick-up.

Remove the joint washer; if broken, this will have to be replaced by a new one on re-assembly.

Use of Detergent Oils

When changing to detergent oils after the car has covered 5,000 miles (8,000 km.) it is of the utmost importance to adhere to the following procedure:—

Flush the sump before filling and fit a new oil filter unit, unless it has recently been changed.

After the car has run a few hundred miles the new oil should be changed, the sump flushed out and a new filter unit again fitted. Subsequently, the normal period between oil changes can be reverted to.

The reason for adopting this procedure is that, with detergent oils, deposits in the engine are released which will collect in the sump and filter, thus promoting clogging.

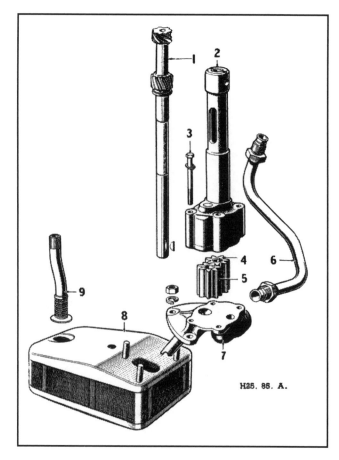

Fig. 5. Details of the oil pump assembly.
1. *Driving spindle.* 2. *Oil pump body.* 3. *Setscrew and washer.*
4. *Driving gear.* 5. *Driven gear.* 5. *Oil delivery pipe.* 7. *Bottom cover.* 8. *Oil strainer.* 9. *Oil return pipe.*

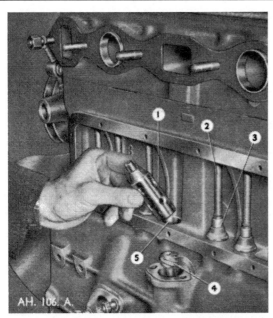

Fig. 6. Tachometer gear and oil pump locating screw.
1. *Gear housing.* 2. *Pinion.* 3. *Washer.* 4. *Bush for pinion.* 5. *Oil pump locating screw.* 6. *Washer.* 7. *Locking cap nut.*

Gauze Strainer and Pick-up

The strainer should be examined for contamination and removed if necessary by releasing three nuts. Wash the gauze with paraffin, using a brush and not a rag.

To Remove the Oil Pump

Disconnect the oil supply pipe from the pump body to the crankcase.

From the left-hand side of the crankcase remove the oil pump locating screw shown in fig. 6. When the locking cap is removed a screwdriver can be used on the screw itself. Note that there is a fibre washer under the nut.

The oil pump complete can now be drawn down out of the crankcase.

To Replace the Oil Pump

Insert the pump from below and push the shaft right home, when the driving gear will mesh with the gear on the camshaft.

Insert the locking screw in the left-hand side of the crankcase and tighten. Fit fibre washer and follow with the cap lock nut.

Replace the oil delivery pipe to the pump body and crankcase.

The pump does not need priming.

To Dismantle and Re-assemble the Pump

The pump body (see fig. 5) is in two pieces; before dismantling, mark the two flanges to assist in re-assembly.

Remove the four long setscrews from the body, and separate the bottom cover and lift out the driven gear.

The driving gear is keyed to the spindle and will need to be tapped off.

Remove the key. The spindle can then be withdrawn from the pump body.

Check that the driving key is in order in the pump spindle and in the keyway of the driven gear.

On re-assembly with the gears in position and the bottom cover bolted up. the pump must be perfectly free from stiffness when rotated.

Release Valve

Release valve pressure is determined by the spring, which is held in position by a plug. This plug is screwed home and no adjustment is possible. (See inset fig. 7).

The valve is conical-faced hollow plunger. Check that the plunger and the valve seat are clean and undamaged and that the passages in the crankcase are clear.

When re-assembling make sure the fibre washer is fitted under the head of the valve plug, and that an oil-tight joint is made.

Valve Rocker Shaft

The valve rocker shaft on the cylinder head is hollow. It is supplied with oil by a pipe connection and is drilled for lubrication of each rocker bearing.

This shaft is plugged at each end, one of these being screwed in and is detachable in order that the shaft may be cleaned internally.

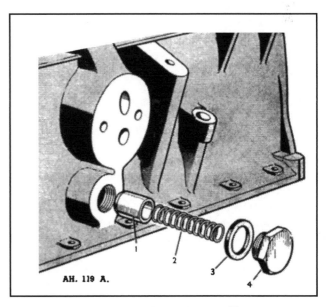

Fig. 7. Oil release valve assembly.
1. *Release valve.* 2. *Spring.* 3. *Washer.* 4. *Cap nut.*

SERVICE OPERATIONS WITH ENGINE IN POSITION

Valve Mechanism

The overhead valve operating mechanism of this engine incorporates in its design an oil cushion to ensure silence of operation, and it is important that a full knowledge of this should be in the possession of those checking or adjusting the valve clearances.

The push rod and ball cup are of one-piece design, the cup being a press fit into the hollow push rod. When checking the valve clearance, it is important to maintain pressure with a screwdriver on the tappet adjusting screw to dispense the films of oil in both the push rod and tappet cups, otherwise a misreading of the clearance may be obtained.

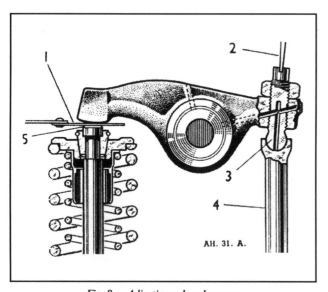

Fig. 8. *Adjusting valve clearance.*
1. *Feeler gauge.* 2. *Screwdriver.* 3. *Adjusting screw.* 4. *Push rod.*
5. *Valve stem.*

Adjusting the Valve Clearance

Lift the valve cover after removing the two securing cap nuts.

Between the solid type of rocker arm head and the valve stem there must be a clearance of .012 in.

To check this adjustment the engine must be turned and the point noted at which the push rod stops falling. From that point until it starts to move again there must be this clearance of .012 in. Test with feeler gauge.

As there is no provision for a starting handle, the engine can be turned by putting the car in top gear and pushing the car slowly forward whilst observing the rise and fall of the push rods. An alternative method of engine turning is achieved by using a spanner on the crankshaft pulley nut.

If adjustment is necessary, whilst continuously applying sufficient pressure to the adjusting screw with a heavy screwdriver slacken the lock nut, raise or lower the adjusting screw in the rocker arm.

Tighten the lock nut when the adjustment is correct, but always check again afterwards in case the adjustment has been disturbed during the locking process.

In replacing the valve cover take care that the joint washer, using a new one if necessary, is properly in place to ensure an oil-tight joint.

Removing the Push Rods

To remove the push rods (see fig. 9) it is not necessary to dismantle the valve gear beyond slackening the tappet adjustment.

Take off the tappet cover.

Slacken the tappet adjustment screw to its full extent. With the aid of a screwdriver, supported under the rocker shaft, depress the valve and spring and then slide the rocker sideways free of the push rod. Remove the push rod. When removing the push rod ensure that the oil cushion is dispersed.

In respect of the front or rear end rocker, however, it is necessary to take out the split cotter pin from the end of the shaft, when the rocker can be removed, together with the plain washer.

Replace in reverse order.

Fig. 9. *Push rod removal.*
1. *Screwdriver.* 2. *Valve spring cup.* 3. *Rocker.* 4. *Spring.* 5. *Push rod.*

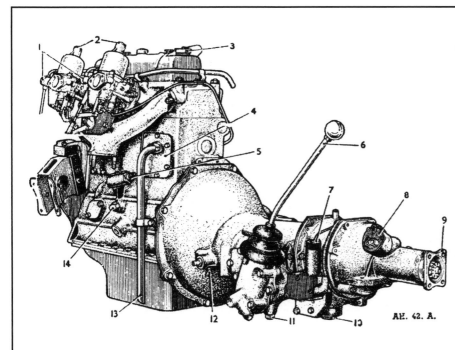

Fig. 10. *Left-hand view of Power Unit.*

1. Carburetter fuel intake.
2. Carburetter hydraulic damper.
3. Oil filler cap.
4. Tappet cover.
5. Tachometer drive.
6. Gear lever.
7. Overdrive solenoid switch.
8. Overdrive governor.
9. Driving flange.
10. Overdrive drain plug.
11. Gearbox drain plug.
12. Clutch cross-shaft oil nipple.
13. Crankcase breather pipe.
14. Oil pump locking screw.

AH. 42. A.

Rocker Arm Bushes

While the rocker gear is detached from the head check for play between the rocker shaft and rocker arm bushes. If this is excessive, new bushes should be fitted. To do this, take out the split pin at the end of the shaft when the plain and spring washers, rocker arm and rocker shaft brackets may be removed.

The white metal bush is best removed using a drift and an anvil. The anvil is recessed to retain the rocker in position while the bush is pressed or gently knocked out with a drift.

The flange of the drift is also recessed to prevent the new split bush from opening when being driven into position. These new bushes are not supplied at a finished size, the internal diameter must be reamed to suit the shaft.

Decarbonising

For this operation it will be necessary to remove the carburetter, manifolds, cylinder head, and push rods.

Scrape off all carbon deposit from the cylinder head and ports (see valve grinding for access to ports).

Clean the carbon from piston crowns by scraping with a tool such as a screwdriver. Care being taken not to damage the piston crowns and not to allow dirt or carbon deposit to enter the cylinder barrels or push rod compartment.

When cleaning the top of the piston, do not scrape right to the edge, as a little carbon left on the chamfered edge assists in keeping down oil consumption; with the pistons cleaned right to the edge, or with new pistons, oil consumption is often slightly, though temporarily, increased.

Removing the Cylinder Head

Drain all water from the cooling system (if the water contains anti-freeze mixture it should be run into a clean container and used again).

Detach the top water hose from the cylinder head and disconnect the heater pipes.

Disconnect the high tension wires from the sparking plugs; also remove the plugs.

Remove the carburetter.

Remove the exhaust and inlet manifolds (see section C, page 1).

Remove the valve rocker cover.

Remove the 11 nuts holding the cylinder head to the cylinder block.

Slacken back the tappet adjusting screws and remove the push rods.

Lift the cylinder head, best accomplished by using a sling under the rocker shaft; a rope is preferable to chains.

With the head on the bench, remove the rocker oil feed pipe (see 9, fig. 11) and then the rocker brackets (note the position of the bracket which carries the oil feed pipe from the cylinder head). This will assist in replacing. There are two holding down nuts to each bracket.

Lift off the rocker gear complete.

Replacing the Cylinder Head

Refit the cylinder head joint washer with the side marked "TOP" uppermost, having smeared both sides with grease to make a good joint and prevent sticking when the head is again lifted.

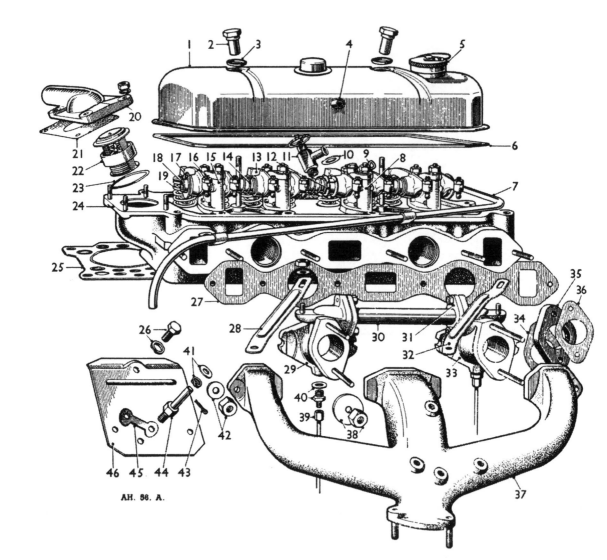

Fig. 11. Cylinder head exploded.

AH. 58. A.

1. *Valve rocker cover.*
2. *Cap nut.*
3. *Washer.*
4. *Breather.*
5. *Oil filler cap.*
6. *Joint washer.*
7. *Heater pipe.*
8. *Rocker shaft bracket.*
9. *Oil pipe.*
10. *Washer.*
11. *Heater valve.*
12. *Distance piece.*
13. *Rocker arm.*
14. *Spring.*
15. *Rocker shaft bracket.*

16. *Adjusting screw.*
17. *Split pin.*
18. *Rocker shaft.*
19. *Valve spring.*
20. *Thermostat cover.*
21. *Joint washer.*
22. *Thermostat.*
23. *Washer.*
24. *Cylinder head.*
25. *Cylinder head washer.*
26. *Setpin and washer.*
27. *Manifold washer.*
28. *Carburetter support.*
29. *Inlet manifold.*
30. *Balance pipe.*
31. *Pipe connection.*

32. *Carburetter support.*
33. *Inlet manifold.*
34. *Joint washer.*
35. *Distance piece.*
36. *Joint washer.*
37. *Exhaust manifold.*
38. *Nut and washer.*
39. *Carburetter drain pipe.*
40. *Drain pipe unions and washer.*
41. *Flat and spring washer.*
42. *Nut and washer.*
43. *Split pin*
44. *Fulcrum pin.*
45. *Lock washer.*
46. *Carburetter shield.*

Replace the rocker gear if this has been removed.

Lower the head over the studs, replace the cylinder head nuts finger tight, and insert the tappet push rods.

Tighten the cylinder head nuts evenly, a quarter of a turn at a time, and in the order shown in fig. 12.

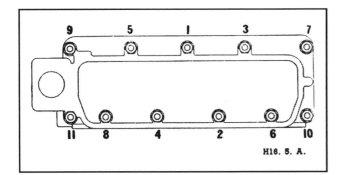

Fig. 12. Tighten the cylinder head nuts in the order illustrated.

Reset the tappets to .012 in. and replace the valve cover.

Replace the manifolds and carburetter making sure that good joints are made, connect up the radiator, and heater pipes, replace the sparking plugs and high tension wires.

Refill the radiator.

Check the valve tappet clearance again after the vehicle has run about 100 miles as the valves have a tendency to bed down. At the same time it is advisable to test the cylinder head nuts for tightness. Tightening the cylinder head nuts may affect tappet clearances, although not usually enough to justify resetting.

Valves

Weak compression in any cylinder, in spite of correct tappet clearances, usually suggest that valve grinding is necessary, and the head should be removed for investigation.

Removing and Refitting a Valve

With cylinder head removed, a valve fitting tool as illustrated in fig. 13, can be used to compress the spring.

Take away the circlip (2) and the split cotters (1) fig. 14, then release the spring and remove the valve.

Re-assembly is a reversal of the operations for removal. When fitting the split cotters it is also worth noting that the spring circlip should be replaced as soon as the cotters are in position. This saves holding the cotters in the groove while the spring is released.

Separate the cotter cup from the oil seal retainer. If the rubber seal shows any signs of damage or perishing, it should be renewed as its object is to prevent excess oil entering the valve guide.

When removing the valves, place them on a valve carrying board to enable them to be identified as to the cylinders from which they have been taken. The valve springs should be tested. The free length being 2⅛ in. for the outer spring and 1⅞ in. for the inner spring; replace if necessary.

Clean the carbon from the top and bottom of the valve heads, as well as any deposit that may have accumulated on the stems. The valve heads should, if necessary, be refaced at an angle of 45°. If the valve seats show signs of excessive pitting it is advisable to reface these also.

The valves now fitted are being made without any indentures or slots in the head, this necessitates the use of a rubber suction headed valve grinding tool. (See fig. 15.)

Valve Grinding

For valve grinding a little grinding paste should be smeared evenly on the valve face, and the valve rotated backwards and forwards against its seat, advancing it a step at short intervals, until a clean and unpitted seating is obtained. The cutting action is facilitated by periodically lifting the valve from its seat. This allows the grinding compound to re-penetrate between the two faces again, after being squeezed out.

On completion, all traces of compound must be removed from the valve and seating.

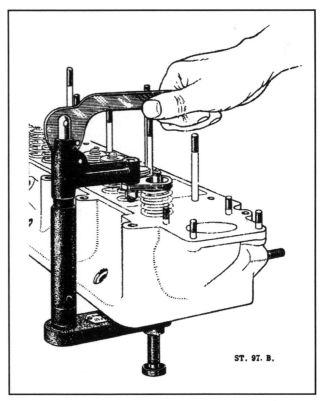

Fig. 13. Showing the valve spring compressor (Service Tool 18G 45) in use.

It is essential for each valve to be ground-in and refitted on its own seating as indicated by the number on the valve head. The valves are numbered 1 to 8, starting from the front. If a new valve is used it should be identified with its seating by stamping the number on the head, taking care not to distort the valve in the process.

It is also desirable to clean the valve guides. This can be done by dipping the valve stem in petrol or paraffin and moving it up and down in the guide until it is free. Reclean the valve and re-insert in the guide, the valve springs, cup, cotters, and circlip being fitted round it.

Renewing Valve Spring in Position

In an emergency a new valve spring can be fitted without lifting the cylinder head, but it is advisable to first bring the piston to top dead centre, to ensure that the valve cannot fall into the cylinder during the process.

Remove the sparking plug and by means of a screwdriver or similar tool, the valve can be held on its seat whilst the springs are compressed. The valve rocker shaft can be used as a fulcrum point by an operator using two screwdrivers to bear on the valve spring cup each side of the valve stem, while the cotters are dealt with.

Distributor. To Remove and Replace

Remove the vacuum control pipe at the union.

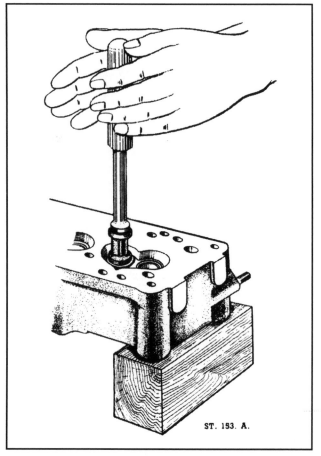

Fig. 15. *The suction type valve grinding tool.*

Slacken the pinch-bolt in the clip at the bottom of the distributor, which can now be turned to allow more easy access to the locating screw which anchors the base plate of the vacuum timing unit on early models. Below the distributor is another set screw, upon removal of which the distributor, complete with driving spindle and spur gear, can be withdrawn. The driving shaft is more readily drawn out if the distributor is given a turning movement while being lifted.

Ignition Timing

The spark should occur on cylinders one and four, $6°$ before "top dead centre" or a $\frac{1}{4}$ in. before the 1/4 mark on the flywheel, as the piston is completing its compression stroke.

In order to reset the ignition timing, first remove the rocker cover; then engage top gear and move the car until No. 1 position is at T.D.C. of the compression stroke, i.e. 7 and 8 valves just rocking. By using a spanner on the crankshaft pulley nut, final location of T.D.C. can be more accurately found.

Remove the distributor cover, slacken the screw in the clip of the distributor casing and turn the casing until the contact breaker points are fully open, with the rotor pointing at No. 1 electrode in the cover. The spark is

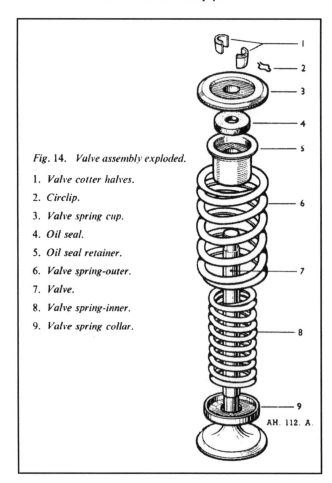

Fig. 14. *Valve assembly exploded.*

1. *Valve cotter halves.*

2. *Circlip.*

3. *Valve spring cup.*

4. *Oil seal.*

5. *Oil seal retainer.*

6. *Valve spring-outer.*

7. *Valve.*

8. *Valve spring-inner.*

9. *Valve spring collar.*

then correctly timed. Finally, lock the distributor in position and refit the cover.

As the distributor cover carries the electrodes for the four cylinders, it is imperative for the rotating arm to pass the spark to the correct sparking plug lead at each firing stroke, therefore the leads must be connected to the sparking plugs so that the firing sequence is 1, 3, 4, 2.

By turning the vernier adjustment knob, there is a considerable latitude for adjustment, but only an extremely small movement should be made at one time. This final adjustment can be made when testing the car on the road.

Sparking Plugs

The sparking plugs fitted to the Austin-Healey are Champion NA8 14 mm. long-reach type.

The gaps of these plugs should be maintained at .025 in. If the gap is allowed to become too wide, misfiring at high speeds is liable to occur; and if too small, bad, slow running and idling will be the result.

Sparking plugs should be regularly inspected cleaned and tested. This is of vital importance to ensure good engine performance, coupled with fuel economy.

When removing the plugs from the engine, use a box spanner, this will avoid possible damage to the insulator, and always remove the copper washers with them. They should then be placed in a suitable holder which has holes drilled to admit the upper end of the plugs and marked to identify each plug with the cylinder from which it has been removed.

The plugs should now be carefully examined and for guidance compared with a new plug.

Oil fouling will be indicated by a wet shiny black deposit on the insulator. This condition is usually caused by worn cylinders, pistons or gummed rings. Oil vapour is forced from the crankcase during the suction stroke of the piston which fouls the plugs.

Petrol fouling will cause a dry fluffy black deposit to be apparent on the plugs. This is usually caused by faulty carburation or ignition system. In this latter connection the distributor coil or leaking and worn out ignition leads may be contributory causes.

Under the above conditions, if the plugs otherwise appear to be sound, they should be thoroughly cleaned, adjusted and tested.

When preparing for cleaning, the plug washers should be removed and examined. The condition of these washers is important, in that a large proportion of the heat from the plug insulator is dissipated to the cylinder head by them. The washer should therefore be reasonably compressed. A loose plug can be easily overheated, thus upsetting its heat range and causing pre-ignition, with consequent short plug life. On the other hand, do not overtighten. All that is needed is a good seal between the cylinder head and the plug. Tightening too much will cause distortion of the washer, with the possibility of blow-by which will again lead to overheating and resulting dangers. If there is any question of defect, replace with new washers.

The plugs should now be thoroughly cleaned of all carbon deposit, resorting to scraping if necessary, removing as much as possible from the space between the insulator and shell. An oily plug should be washed out with petrol. If a plug cleaning machine is available, 5 to 10 seconds in this will remove all remaining signs of carbon. Remember to thoroughly "blow out" the plug after treatment under these conditions, in order to remove all traces of abrasive left inside.

After cleaning, thoroughly examine the plug for cracked insulator or worn away insulator nose. Should either of these conditions be apparent a new plug should be installed.

Carbon deposit on the threads of the plug should be carefully removed by using a wire brush or, if available, a wire buffing wheel. Take care not to damage the electrodes or insulator tip. This often neglected cleaning operation will lead to tight threads and resultant loss of heat dissipation due to the carbon deposit, and thereby cause overheating.

The condition of the electrodes should now be noted and any signs of corrosion removed, if it is felt that the plugs are worthy of further use. This can be carried out with the use of a small file, to carefully dress the gap area. The gap should then be reset, using the plug gauge

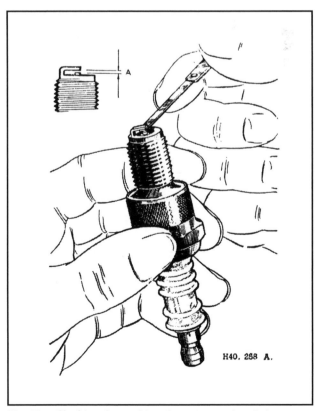

H40. 268 A.

Fig. 16. Checking the sparking plug gap A with a feeler gauge.

provided in the tool kit. When resetting, bend the side electrodes only, never bend the central one as this may split the insulator tip.

If a plug testing machine is available, the plugs can be accurately tested to ascertain their fitness for further service in the car. The gaps should be set to .022 in. before being subjected to this test. A plug can be considered fit for further use if it gives a continuous spark, when in the testing machine, up to 100 lbs. per square inch. Having been found satisfactory, the gaps should be reset to .025 in. before refitting to the engine.

It is advisable, whilst the plugs are under pressure in the testing machine, to apply a spot of oil to the terminal end, to check for air leakage. Excessive leakage here will

tend to cause compression loss, rapid deterioration of the electrode and overheating of the electrode tip. The top half of the insulator should also be carefully examined for any signs of paint splashes or accumulations of grime and dust, which should be removed. Should there be any signs of cracks due to faulty use of the spanner, the plug should be replaced.

Renewal of the sparking plugs is left to the owner's discretion as their efficient life is variable. When replacing the plug lead, make sure that it is securely attached.

Make plug inspection, cleaning and testing a routine job and carry this out at least every 6,000 miles. Remember, plugs in good condition will ensure better fuel consumption and good engine performance.

REMOVING AND REFITTING THE ENGINE

When removing the engine it is advised that the engine, gearbox and overdrive be removed as one complete unit. If this method is adopted the refitting procedure will be appreciably simplified, although the actual removing and replacing of the unit will require two operators.

Drain the water from the cooling system and ascertain that the battery master switch is in the off position. The engine, gearbox and overdrive may be drained of oil, but this is not an essential operation.

The first of the dismantling operations is to remove the bonnet, procedure described in the "Bodywork" section of the manual. The radiator should also be removed, see page B/4.

Fig. 18. *The engine left-hand front mounting bracket showing the four setpin holes at A.*

Working from the left-hand side of the engine remove the setpins securing each air cleaner to its carburetter flange. Release the four setpins holding the engine mounting bracket to the chassis, (see figs. 17 and 18). Disconnect the choke cable at the point where it joins the carburetter linkage. The throttle is best disconnected at the ball joint connection adjacent to the rear end of the exhaust manifold. Slacken the clips securing the water return hose at the rear end of the valve rocker cover.

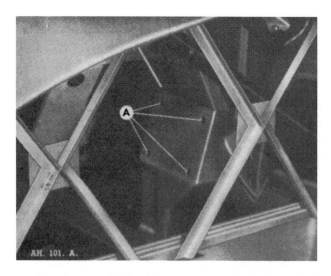

Fig. 17 *The engine left-hand front mounting point showing at A the four tapped holes.*

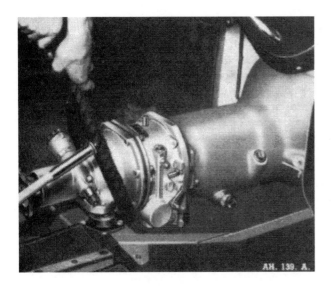

Fig. 19. Using a rope to lift the rear end of the unit to clear the chassis.

Now, working from the right-hand side of the engine remove the four engine mounting bracket setpins. Release the oil filter which is held to the crankcase by two setpins and washers. Disconnect the lead from the starter then the oil gauge flexible pipe at its cylinder block union. Release wires from the dynamo, distributor and coil. The last operation on this side is to slacken the clip securing the heater flexible pipe to the control valve mounted on the cylinder head.

Operating inside the car remove both seat cushions and the central armrest together with the short piece of propeller shaft tunnel. Unscrew the gear lever knob locknut and remove both knob and locknut. Release

Fig. 20. Illustrating a specially designed bracket to give the correct lifting angle.

the seven screws holding the gearbox cover and the six screws securing the bridge piece bulkhead cover. Remove both covers.

To facilitate the removal of the four bolts with nuts and lock washers from the universal joint, push the car backwards or forwards in order to rotate the propeller shaft.

Disconnect the wires from the switch on the right-hand side of the gearbox; also the wires from the governor and solenoid on the overdrive unit. These wires are secured to the gearbox by two clips—one on the side cover and the other on the adapter plate. Release the clips and secure the wires out of harm's way. The gear lever should next be removed. With the three nuts and washers inside the gear lever cover released, the lever together with its cover, rubber packing pad, seating ring and rubber sealing washer can then be withdrawn. Take

Fig. 21. Showing the power unit at the correct lifting angle using a different method to that in Fig. 20.

care not to lose the three small distance pieces that fit over the studs.

Disconnect the tachometer cable from its driving point on the left-hand side of the cylinder block. The speedometer cable should be freed from the right-angled drive on the right-hand side of the overdrive. Unscrew the nut holding the clutch lever to the cross-shaft and remove the lever. Remember to unhook the clutch linkage return spring from the bell housing.

The last operation inside the car is to release both rear mounting setpins securing the overdrive to the chassis.

The rest of the dismantling operations are dealt with from beneath the car. Therefore, if there is no pit or hydraulic ramp available the car should be jacked up and suitable trestles placed under all four wheels.

Disconnect the short flexible petrol pipe at its upper

union situated at the engine mounting bracket. Release the exhaust down-pipe steady bracket from the crankcase and from the down-pipe. Now disconnect the down-pipe from the exhaust manifold flange. A long "T" spanner will be found suitable for unscrewing the three securing nuts and washers. The short earth lead from the starter to the chassis should be released at the chassis. Finally, extract the gearbox stabiliser as described on page F/3 of gearbox removal.

A suitable lifting bracket (see figs. 20 and 21) should now be secured in position, irrespective of the type of lifting tackle used, the final unit lifting angle should be approximately 45° thus allowing it to clear the surrounding bodywork.

As previously mentioned, the actual lifting operation will require two operators; one to lift the rear end of the unit (see fig. 19) whilst the other operates the lifting tackle and guides the unit through the bonnet opening.

Refitting is a reversal of the removal procedure.

OPERATIONS WITH THE ENGINE REMOVED

The following operations should be carried out with the engine removed, although in some cases it is possible to perform them with the engine in position.

Before removing or replacing any component it is important to ensure that all surrounding surfaces are perfectly clean to prevent the entry of foreign matter into any vital parts. This can best be accomplished by the use of a paraffin bath and a brush, and it is also important to note that fluffy rags should never be used as there is a danger of causing obstruction to small oil ways.

Valve Guides

The valve guides are of a one-piece design. They are pressed into the cylinder head with $\frac{11}{16}$ in. of the guide protruding above the head.

Internally only the exhaust valve guides are stepped to give three different diameters. The portion of the valve guide above the cylinder head has the closer fit around the valve stem. This is due to the freedom for expansion afforded this part of the guide; the close fit thus restricts the passage of oil and gas along the valve stem.

The portion of the guide sunk into the head cannot have so close a fit to the valve, due to the unequal expansion of the gas heated valve stem and the indirect water cooling of the guide.

To position each inner valve spring at the cylinder head, a collar (see fig. 14) is fitted over the part of the valve guide protruding from the cylinder head. The outer valve spring fits into a recess in the cylinder head.

Valve guides should be tested for wear whenever valves are removed, and if excessive side play is present, a close check should be made of the valve stem and the guide. In the event of wear being noticeable, the defective component should be renewed. If a valve is at fault the wear will usually be evident on the stem, but it should be borne in mind that the valve and stem should be a running fit to avoid the possibility of an air leak.

If renewal is necessary due to wear the valve guide may be driven out after the removal of the valve as shown in fig. 22.

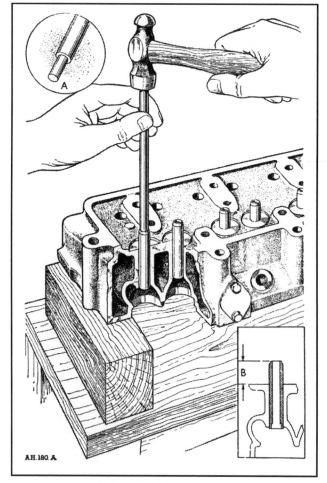

Fig. 22. Renewing a valve guide.
A. Stepped drift. B. Final position of guide above cylinder head (11/16 in.).

The drift is shown stepped in order to ensure location and to obviate it slipping off the guide and damaging the port. The guide should be knocked out in the direction shown, working on the plain and not the chamfered end.

A new guide should be forced into position in the same direction—that is, inserting it through the valve seating and driving towards the top of the cylinder head.

The final position of the guide is shown in fig. 22. It must stand $\frac{1}{16}$ in. above the cylinder head.

Removing the Tappets

Remove the valve rocker cover, slacken back the tappet adjustment and withdraw the push rods.

Remove the tappet cover—on the left-hand side of the engine; it is held in place by 15 set screws, with fibre washers. Carefully remove the joint washer. Renew, if it is damaged.

Remove the vent pipe attached to the cover; it is flange mounted with two set screws.

The tappet plungers may be withdrawn from the crankcase by lifting upwards with finger and thumb (fig. 24).

Replacement is a reversal of this operation, but take care to make an oil-tight joint with the cover and the vent pipe.

VALVE TIMING GEAR

For access to the valve timing gears or chain, first drain and remove the radiator complete (see section B, page 4).

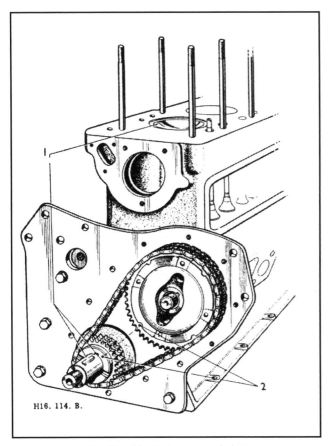

Fig. 23. *The timing gear and chain.*
1. *Pistons 1 and 4 at T.D.C. with the relative position of the crankshaft keys.* 2. *The correct positions of the spots on both timing gears.*

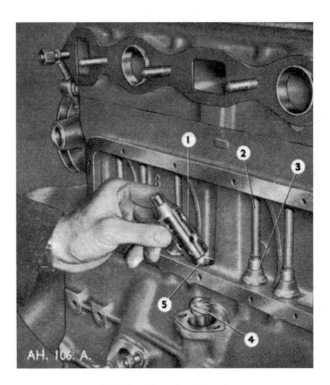

Fig. 24. *Removing a tappet.*
1. *Tappet.*   2. *Push rod.*   3. *Tappet cup.*   4. *Oil pump spindle.*
5. *Tappet cylinder.*

Starting Nut and Belt Pulley

Using a suitable heavy box spanner, unscrew the starting nut, see 26, fig. 30, on the crankshaft after knocking back the lock washer. The spanner will probably have to be hammered in order to "start" this nut, but a few fairly sharp blows in an anti-clockwise direction should be sufficient.

With the nut and its washer removed, the belt pulley can be withdrawn from the shaft. The pulley is keyed but there is no taper fit and withdrawal will present no difficulty, but an extractor can be used to advantage.

Timing Cover

The timing gear cover (see 2, fig. 30) is held to the engine by set screws. These are of two sizes, $\frac{7}{16}$ in. and $\frac{1}{4}$ in.

There are special oval-shaped washers and spring washers under each screw head.

The larger set screws—seven in number—are used in the lower portion of the cover, where the holes are in the crankcase; the smaller set screws—five in number— enter the engine front mounting plate only.

The timing cover and paper washer can now be removed and at the same time the oil thrower (see 23, fig. 30) should be taken from the front of the crankshaft. Note the correct fitting to prevent oil from creeping to the fan pulley; the concave or hollow side must face front towards the pulley.

When re-assembling do not damage the felt washer, make the joint carefully, using a new joint washer if necessary, and tighten set screws evenly.

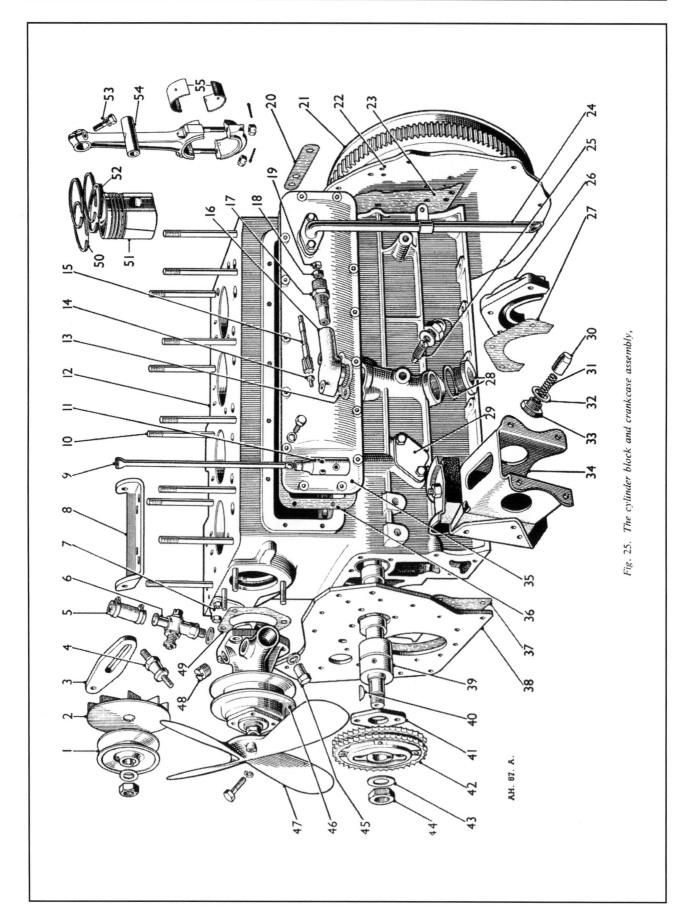

Fig. 25. The cylinder block and crankcase assembly,

AH. 67. A.

Caption for Fig. 25

1. Dynamo pulley.	19. Oil seal retainer.	38. Front mounting plate.
2. Dynamo fan.	20. Joint washer.	39. Camshaft.
3. Adjusting link.	21. Flywheel.	40. Key.
4. Adjusting link stud.	22. Rear mounting plate.	41. Camshaft locating plate.
5. Heater pipe connection.	23. Joint washer.	42. Camshaft gear.
6. Shut-off valve.	24. Crankcase breather pipe.	43. Lockwasher.
7. Front cover for water gallery.	25. Oil pump locating screw.	44. Nut.
8. Dynamo mounting bracket.	26. Rear cover.	45. Extension nut.
9. Push rod.	27. Joint washer.	46. Fan and pump pulley.
10. Cylinder head stud.	28. Plug and washer.	47. Fan.
11. Tappet.	29. Blanking plate.	48. Plug.
12. Cylinder block.	30. Oil release valve.	49. Joint washer.
13. Joint washer.	31. Spring.	50. Compression ring.
14. Thrust button.	32. Washer.	51. Piston.
15. Pinion.	33. Cap nut.	52. Oil control ring.
16. Gear housing.	34. Mounting bracket.	53. Clamping bolt.
17. Bush for pinion.	35. Side cover.	54. Gudgeon pin.
18. Oil seal.	36. Joint washer.	55. Bearings.
	37. Joint washer.	

Front Mounting Plate

The engine front mounting plate can be removed by taking out the remaining set screws.

Removal of the Oil Sump (see page D/9).

Removal of the Flywheel

After taking away the clutch (see Section E/3), the flywheel can be removed upon releasing the four bolts to crankshaft flange.

In replacing the flywheel, see that the 1/4 timing mark is in line with the first and fourth throws of the crankshaft.

Rear Mounting Plate

The engine rear mounting plate may be removed after the flywheel by taking out the remaining set screws into the crankcase.

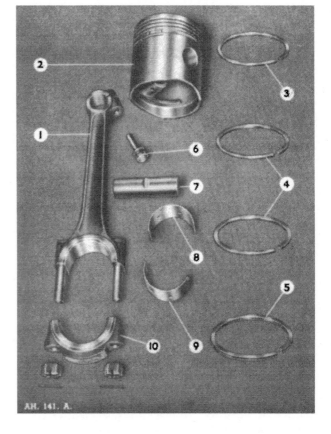

Fig. 27. Piston and connecting rod assembly.

1. Connecting rod.	6. Clamping bolt.
2. Piston.	7. Gudgeon pin.
3. Compression ring (plain).	8. Bearing.
4. Compression rings (taper).	9. Bearing.
5. Oil control ring.	10. Connecting rod clamp.

Pistons and Connecting Rods

There are two oil jets in the top half of the big end bearing. Ensure these line up with holes in the shell bearings to give free passage of oil.

For drawing the pistons, the connecting rods have to be taken upwards through the cylinder bores; sump and cylinder head have therefore to be removed.

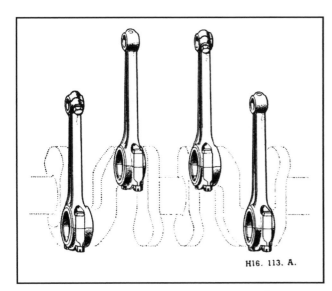

Fig. 26. Order of assembling the connecting rods. Note the alternative positioning of the small ends.

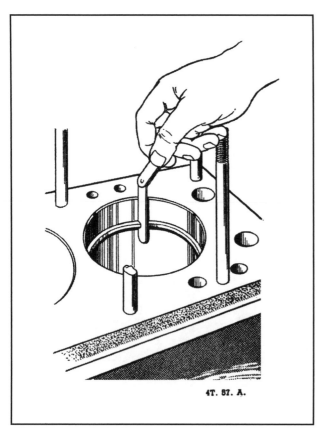

Fig. 28. *Measuring piston ring gap.*

Remove the split pins and nuts from the big end bearing cap and withdraw the cap. Before pushing the connecting rod through the bore check that the bearing bolts are still properly in position.

The heads of the bolts are anchored in the top half of the cap by a small peg. If these have been disturbed the big end may not pass through the bore as the head may have turned.

See page D/3 for the bearing sizes and fits. Check the crank for out-of-round and scoring; if either is present the crankshaft will have to be removed for grinding.

The shell bearings are removed by hand and new bearings require no scraping or hand fitting, apart from placing in position so that the feathered ends are properly located on top and lower halves (fig. 27).

When replacing the split pins in the big end and main bearing nuts, after tightening up, the ends should be bent back with pliers; hammering back the cotters is not approved.

Gudgeon Pin Clamping Bolt

The gudgeon pin clamping bolt is set at an angle and a spring washer is used. As will be seen from fig. 26 the rods should be fitted facing alternately.

Removing a Piston

Slacken the clamping screw in the small end of the connecting rod to release the gudgeon pin and remove the piston. The gudgeon pin is a push fit in the piston.

Care should be taken in withdrawing the connecting rod and piston, or damage may be caused to the piston or the rings.

Pistons and Bores

Piston clearance is measured at the skirt, measurement (see D/2) being taken at right-angles to the gudgeon pin and in the working part of the cylinder bore. (See fig. 23). Note that the pistons are of split skirt design. Piston ring gaps (see D/2) should be tested in the cylinder bores. Ensure that the rings are free in their respective grooves in the piston.

A piston ring guide will facilitate the replacement of the piston assembly.

Oversize Piston Rings

After fitting oversize piston rings there may be a tendency to noisy operation unless attention is paid to any bore "lip" which may be present in the cylinder. A dial gauge in the cylinder will show the presence of such a "lip"; it should be eased with the aid of a hand scraper.

Bearings

These bearings can often be dealt with while the engine is in the chassis, but for major overhauls it is preferable to take out the engine unit complete. These bearings are most easily serviced with the engine inverted.

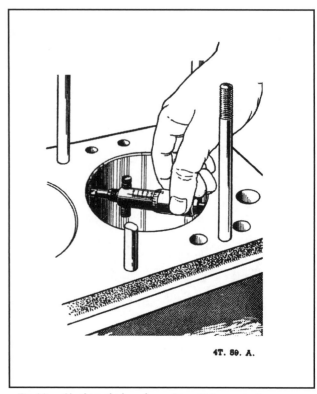

Fig. 29. *Checking the bore for ovality with internal micrometer.*

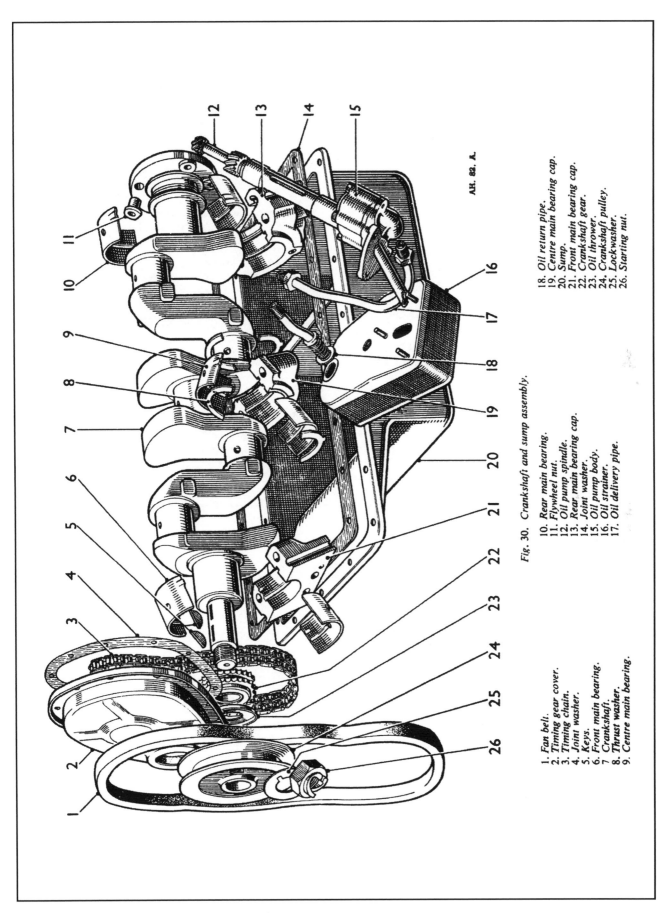

A.H. 82. A.

Fig. 30. Crankshaft and sump assembly.

1. Fan belt.
2. Timing gear cover.
3. Timing chain.
4. Joint washer.
5. Keys.
6. Front main bearing.
7. Crankshaft.
8. Thrust washer.
9. Centre main bearing.
10. Rear main bearing.
11. Flywheel nut.
12. Oil pump spindle.
13. Rear main bearing cap.
14. Joint washer.
15. Oil pump body.
16. Oil strainer.
17. Oil delivery pipe.
18. Oil return pipe.
19. Centre main bearing cap.
20. Sump.
21. Front main bearing cap.
22. Crankshaft gear.
23. Oil thrower.
24. Crankshaft pulley.
25. Lockwasher.
26. Starting nut.

Centre Bearing

There are thrust washers fitted on each side of the centre bearing. See that the peg formed on each pair fits into the bearing cap. The end float permissible is from .002 to .003 in.

Bearing Caps

The front and rear main bearing caps have cork oil sealing plugs fitting into a drilling on each side (fig. 31). In rebuilding see that these plugs are in place and in good condition.

Provided the journals are not unduly worn new shell bearings can be fitted a pair at a time, simply by removing the bearing caps and exchanging the shells.

Handle the new shell bearing halves carefully as they have a very fine finish, and ensure that all dirt and grit is removed from the bearing caps and the journal faces.

When fitting bearings ensure that all bearing caps are replaced the right way round as shown by the stamp markings which face the camshaft. See that connecting rod caps are retained for the same connecting rods, and that they are refitted the same way round as found when dismantling.

Fig. 31. Showing the front main bearing cap with the cork sealing plugs at 1 and 2.

Timing Chain Tensioner

The rubber tensioner ring (see Section D/8, fig. 4) fitted to the camshaft gear sprocket ensures quiet running by constantly taking up the slack in the timing chain.

If rattles are diagnosed as coming from the timing chain, the chain alone should not be suspected. The rubber tensioner may have deteriorated and if this is found to be so, the tensioner ring must be replaced.

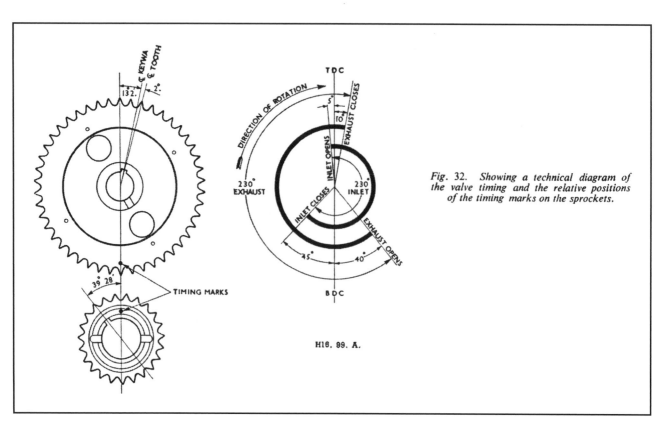

Fig. 32. Showing a technical diagram of the valve timing and the relative positions of the timing marks on the sprockets.

Withdrawing the Camshaft

First remove the oil sump, the oil strainer and the oil pump as described on pages D/9 and 10, followed by the distributor and driving spindle, see page D/15.

The timing gear cover should then be removed together with the valve rocker cover and engine side cover. The push rods, tappets, timing gears and chain can now be taken away.

Unscrew the set screws and spring washers holding the locating plate to the crankcase, and draw the camshaft forward, rotating slowly to assist withdrawal.

There should be a clearance of .003–.006 in. between the camshaft gear and the camshaft shoulder when assembled, to provide a float. Check this with a feeler gauge.

Camshaft Bearings

These can only be renewed with the engine out of the frame, as the engine rear mounting plate must be removed for access to the back bearing.

Old bearing liners can be punched out and new ones tapped into position. Oil holes must be carefully lined up.

All bearings must, however, be reamed in line to give .001 in. to .002 in. clearance on each.

Replacing the Camshaft

Replacement of the camshaft is a reversal of the above procedure.

Fitting a Timing Chain

The chain must be fitted to the sprockets while both are away from the engine, as no spring link is provided.

If the camshaft sprocket has been separated from the camshaft the smaller sprocket must first be engaged with the crankshaft and passed over the keys (see 5, fig. 30). The larger sprocket at the same time passing over the camshaft key and being finally secured by means of a nut and spring washer.

Timing Marks

The timing gears are spot-marked and the timing is correct when the spot on the camshaft gear is in line with the spot on the crankshaft gear at their closest position (see fig. 32). To check the valve timing the inlet valve must be timed to open 5° before top dead centre, the equivalent of which on the flywheel is a point $\frac{9}{16}$ in. before T.D.C. on a diameter of $12\frac{11}{16}$ in., tappets being set to .021 in. before timing. These should be re-adjusted to .012 in. afterwards for normal running.

SERVICE DIAGNOSIS GUIDE

Symptom	No.	Possible Fault
(a) Will Not Start	1	Defective coil
	2	Faulty condenser
	3	Dirty, pitted or incorrectly set contact breaker points (pages O/2 and 3)
	4	Ignition wires loose or leaking
	5	Water on sparking plugs leads
	6	Corrosion of terminals or discharged battery
	7	Faulty starter (page O/8)
	8	Wrongly connected plug leads
	9	Vapour lock in fuel line
	10	Defective fuel pump (page C/10)
	11	Over-choking
	12	Under-choking
	13	Choked petrol filter (page C/12)
	14	Valves leaking
	15	Sticking valves (see under "j")
	16	Valve timing incorrect (page D/20)
	17	Ignition timing incorrect (page D/15)
(b) Engine Stalls		In (a), check 1, 2, 3, 4, 10, 11, 12, 13, 14 and 15
	1	Plugs defective or incorrect gap (page D/16)
	2	Retarded ignition (page D/15)
	3	Mixture too weak
	4	Water in fuel system
	5	Petrol tank breather choked
	6	Incorrect valve clearance (page D/11)
(c) Poor Idling		In (b), check 1 and 6
	1	Air leak at manifold joints
	2	Incorrect slow running adjustment (section C/10)
	3	Air leak in carburetter
	4	Over-rich mixture
	5	Worn piston rings
	6	Worn valve stems or guides
	7	Weak exhaust valve springs
(d) Misfiring		In (a), check 1, 2, 3, 4, 5, 8, 10, 13, 14, 15, 16 and 17
		In (b), check 1, 2, 3 and 6
	1	Weak or broken valve springs
(e) Overheating		See section B, page 2
(f) Low Compression		In (a), check 14 and 15
		In (c), check 5 and 6
		In (d), check 1
	1	Worn piston ring grooves (page D/2)
	2	Scored or worn cylinder bores

Symptom	No.	Possible Fault
(g) Lack of Power		In (a), check 3, 10, 11, 13, 14, 15 and 16
		In (b), check 1, 2, 3, and 6
		In (c), check 5 and 6
		In (d), check 1
		Check (e) and (f)
	1	Leaking joint washers
	2	Fouled sparking plugs
	3	Automatic advance not functioning
(h) Burnt Valves or Seats		In (a), check 14 and 15
		In (b), check 6
		In (d), check 1
		Check (e)
	1	Excessive carbon around valve seat and head (page D/12)
(j) Sticking Valves		In (d), check 1
	1	Bent valve stem
	2	Scored valve stem or guide
	3	Incorrect valve clearance (page D/11)
(k) Excessive Cylinder Wear		Check 11 in (a)
		Check (e)
	1	Lack of oil (page D/7)
	2	Dirty oil
	3	Gummed up or broken piston rings
	4	Badly fitting piston rings
	5	Misalignment of connecting rods
(l) Excessive Oil Consumption		In (c), check 5 and 6
		Check (k)
	1	Ring gap too wide (page D/2)
	2	Oil return holes in piston choked with carbon
	3	Scored cylinders
	4	Oil level too high (page D/7)
	5	External oil leaks
	6	Ineffective valve oil seal
(m) Crankshaft and Connecting Rod Bearing Failure		In (k), check 1
	1	Restricted oilways
	2	Worn journals on crankpins
	3	Loose bearing caps
	4	Extremely low oil pressure
	5	Bent connecting rod
(n) Internal Water Leakage		See section B, page 2
(o) Poor Circulation		See section B, pages 2 and 3

Symptom	No.	Possible Fault
(p) Corrosion		See section B, page 3
(q) High Fuel Consumption		See section C, page 13 (a)
(r) Engine Vibration	1 2 3 4 5	Loose dynamo bolts Fan blades out of balance Incorrect clearance for engine front mounting rubbers Exhaust pipe mountings too tight Incorrect adjustment of stabiliser

CLUTCH

GENERAL DATA

Make	Borg & Beck
Type	Single Dry Plate
Diameter	9 in. (23 cm.)
Total Frictional Area ...	36.8 sq. in. (238 sq. cm.) × 2
Lining Thickness ...	.145–.155 in. (3.683–3.937 mm.)
Withdrawal Bearing ...	Self Lubricating Carbon ring
Number of Springs	9
Total Axial Spring	
Pressure	1,215–1,305 lb. (551–592 kg.)
Distance, Thrust Race	
to Thrust Plate ...	0.06 in. (1.52 mm.) minimum
Thrust Plate Travel to Fully	
Released Position ...	.42–.47 in. (10.6–11.9 mm.)
Pedal Clearance (free movement) ...	$\frac{11}{16}$ in. (17.46 mm.)

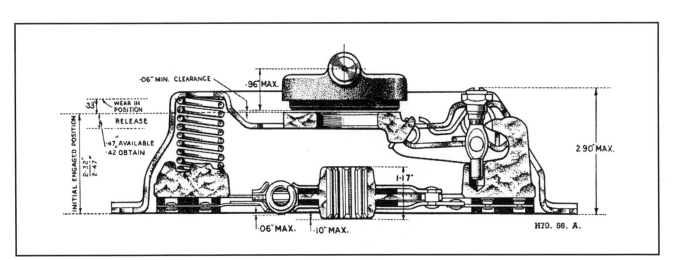

Fig. 1.   Dimensional view of clutch.

DESCRIPTION AND ADJUSTMENTS

Driven Plate Assembly

This is of the flexible type (17) in which the splined hub is indirectly attached to a disc (see Fig. 2), which transmits the power and the over-run through a number of coil springs held in position by retaining wires. Two friction linings are riveted to the disc.

Cover Assembly

The cover assembly consists of a pressed steel cover (8) and a cast iron pressure plate (16) located by nine thrust springs (19). Mounted on the pressure plate are three release levers (13) which pivot on floating pins (11) retained by eye bolts (10). Adjustment nuts (9) are screwed on to the eye bolts and secured by staking. Struts (15) are interposed between the lugs on the pressure plate and the outer ends of the release levers. Anti-rattle springs (14) load the release levers and retainer springs (12) connect the release lever plate.

Release Bearing

The release bearing consists of a graphite bearing shrunk into a bearing cup (7), the cup being located by the operating forks (5) and the release bearing retainer springs (6).

Running Adjustments

The only adjustments necessary throughout the life of the driven plate linings is to restore, periodically, the free movement of the clutch pedal, i.e. movement of the pedal before the release bearing comes into contact with the release lever plate and commences to withdraw the clutch. As the driven plates wear, the free movement of the pedal will gradually decrease, thus tending to prevent the clutch fully engaging and permitting too great a movement on withdrawal. When this free movement becomes reduced, adjustment must be made to restore it to the correct amount, which is $\frac{11}{16}$ in. (17.46 mm.).

CLUTCH ASSEMBLY EXPLODED

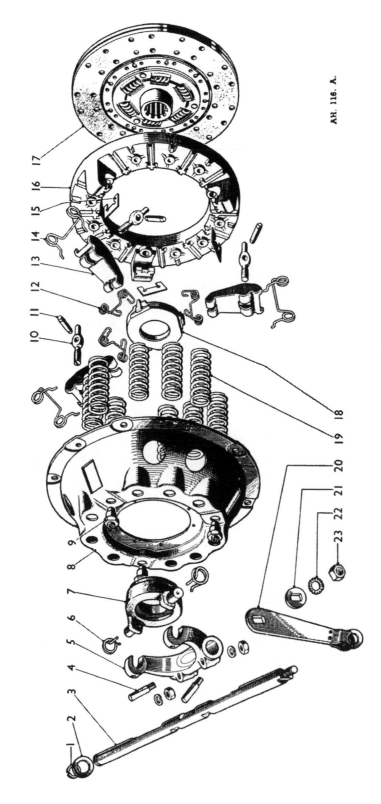

AH. 116. A.

Fig. 2. Components of the clutch.

1. Shaft circlip.
2. Washer.
3. Operating shaft.
4. Withdrawal fork cotter.
5. Withdrawal fork.
6. Release bearing retaining spring.

7. Release bearing and cup.
8. Clutch cover.
9. Eyebolt nut.
10. Eyebolt.
11. Release lever pin.
12. Retainer spring.

13. Release lever.
14. Anti-rattle spring.
15. Strut for release lever.
16. Pressure plate.
17. Clutch plate with linings.
18. Release lever plate.

19. Thrust spring.
20. Operating lever.
21. Operating lever washer.
22. Shakeproof washer.
23. Operating shaft nut.

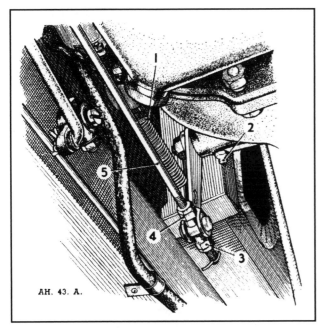

AH. 43. A.

Fig. 3. Clutch pedal adjustment.

1. *Return spring.*   2. *Operating shaft lubricator.*
3 *and* 4. *Adjusting nuts.*   5. *Adjustment rod.*

The free movement for either right or left-hand drive models is easily adjusted by altering the effective length of the rod connected to clutch operating shaft lever. This free movement is increased by first slackening the two locknuts (outer nuts see fig. 3), and then turning the two adjusting nuts (inner nuts) so that the effective length of the rod is increased.

In all cases the adjustment must be such as to allow this free movement to be felt by the pressure of one finger on the clutch pedal. After taking up the free move-

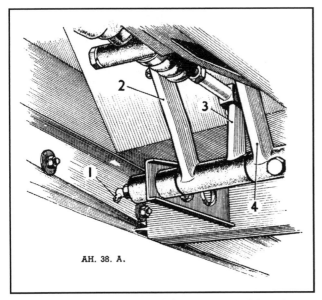

AH. 38. A.

Fig. 4.   Brake and clutch pedal lubrication is served through one nipple (1).   2. Brake pedal.   3. Master cylinder lever. 4. Clutch pedal stalk.

ment the stronger resistance of the clutch springs will be obvious and thus it is easy to ascertain the amount of free movement.

The $\frac{11}{16}$ in. of free movement in the pedal will give a minimum clearance of .06 in. between the graphite release bearing and the release lever plate, thus preventing continual rubbing of the release bearing on the plate.

Removing the Clutch

To gain access to the clutch it is first necessary to remove the gearbox and overdrive complete from the engine (see section F). Before the gearbox and overdrive

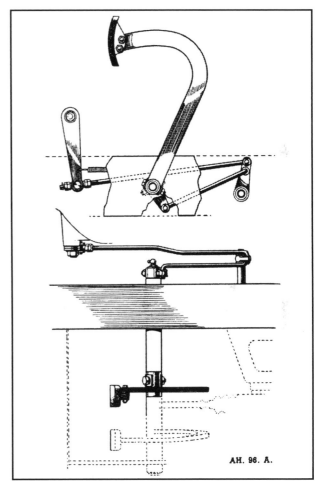

AH. 96. A.

Fig. 5. Clutch pedal linkage layout.

are released from the engine, support the engine at its after end by packing up with suitable wooden blocks or a jack.

Once the gearbox and overdrive unit is free, slacken the clutch cover securing screws a turn at a time by diagonal selection until the spring pressure is relieved. Then remove the screws completely and lift the clutch assembly away from the flywheel. Finally, remove the driven plate assembly.

NOTE: The release levers are correctly set on assembly. Interference with this setting, unless new parts have to be fitted, will throw the pressure plate out, causing judder.

DISMANTLING, ASSEMBLING AND GAUGING THE CLUTCH

By using Service Tool 18G 99, the clutch can be quickly dismantled, reassembled and adjusted to a high degree of accuracy.

The tool comprises the following parts: base plate centre pillar, spacing washers, distance pieces, height finger, actuating mechanism, setscrews, speed brace and metal box. As this tool is universal, a chart indicating the particular parts to be used for particular types of clutch will be found on the inside of the lid of the box.

Dismantling

With a 9 in. clutch, select three spacing washers (2) Fig. 6 and place them over the code letter (D) on the base plate. See Fig. 7.

Now place the clutch on the three spring washers so that the holes in the cover coincide with the tapped holes in the plate, insert the setscrews provided and tighten them, a little at a time, by diagonal selection until the cover is firmly attached to the base plate at all possible points. This is most important if the best results are to be achieved.

Mark the cover, pressure plate lugs and release levers with a centre punch so that the parts can be reassembled in their relative position in order to maintain the balance of the clutch.

Detach the release lever plate from the retaining springs and remove the three eye-bolt nuts or adjusting nuts.

Slowly release the pressure on the springs, unscrewing by diagonal selection, the setscrews securing the cover to the base plate. The clutch cover can then be lifted to expose all components for inspection.

The release levers, eye-bolts, struts and springs should be examined for wear and distortion. Renew these parts as necessary, bearing in mind that the thrust springs must only be renewed in sets.

Clean all parts and lubricate the bearing surfaces of the levers, eye-bolts, etc., sparingly with grease.

Assembling

Place the pressure plate over the three spacing washers on the base plate (12), with the thrust springs (7) in position on the pressure plate (9) see Fig. 7.

Assemble the release lever, eye-bolt and pin holding the threaded end of the eye-bolt and the inner end of the lever as close together as possible. With the other hand, insert the strut in the slots on the pressure plate lug sufficiently to allow the plain end of the eye-bolt to be inserted into the hole in the pressure plate.

Move the strut upwards into the slot in the pressure plate lug and over the ridge on the short end of the lever and drop it into the groove formed in the latter. Fit the other two levers in a similar manner.

Place the cover (4) over the assembled parts, ensuring that the anti-rattle springs are in position, and that the tops of thrust springs (7) are directly under the seats in the cover. In addition the machined portions of the pressure plate lugs must be directly under the slots in the cover through which they have to pass.

Compress the pressure springs by screwing down the cover (4) to the base plate (12) by using the special setscrew (10) placed through each hole in the cover. Tighten the screws, a little at a time, by diagonal selection to prevent distortion of the cover. The eye-bolts (5) and pressure plate lugs must be guided through the holes in the cover at the same time.

Gauging

Screw the nuts (6) in to the eye-bolts and proceed as follows:—

Screw the centre pillar (3) into the base plate and slip the distance piece (2)—code 7 for 9 in. clutch—over

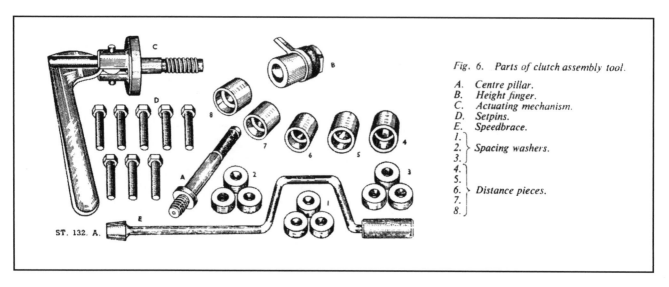

Fig. 6. Parts of clutch assembly tool.

A. Centre pillar.
B. Height finger.
C. Actuating mechanism.
D. Setpins.
E. Speedbrace.
1.
2. } Spacing washers.
3.
4.
5.
6. } Distance pieces.
7.
8.

ST. 132. A.

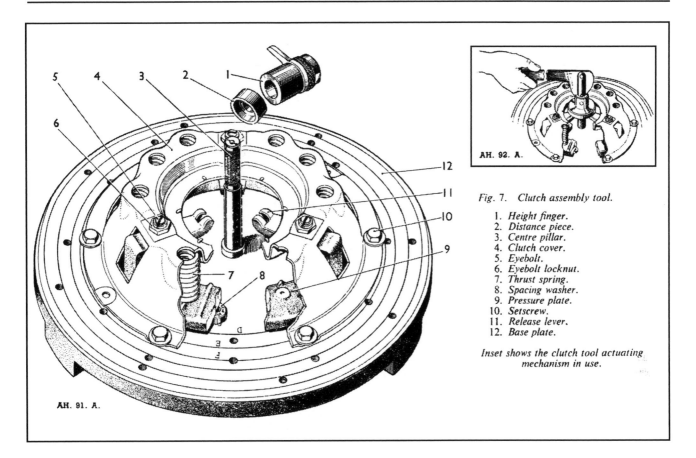

AH. 92. A.

Fig. 7. Clutch assembly tool.

1. Height finger.
2. Distance piece.
3. Centre pillar.
4. Clutch cover.
5. Eyebolt.
6. Eyebolt locknut.
7. Thrust spring.
8. Spacing washer.
9. Pressure plate.
10. Setscrew.
11. Release lever.
12. Base plate.

Inset shows the clutch tool actuating mechanism in use.

AH. 91. A.

the pillar followed by the cam-shaped height finger (1). Adjust the height of the release levers by screwing or unscrewing the eye-bolt nuts until the height finger, when rotated, just contacts the highest point on the tip of the release levers (11).

Replace the height finger and pillar by the clutch actuating mechanism (see inset Fig. 7) and actuate the clutch several times by operating the handle. This will enable the parts to settle down on their knife edges. Replace the height finger and distance piece and readjust the height of the release levers. Finally repeat the procedure to make quite sure the release levers are seating properly and gauge again.

Secure the eye-bolt nuts (6) and fit the release lever plate on the tips of the release levers (11), then secure by means of the three retaining springs.

Release the setscrews (10), a little at a time, by diagonal selection, and remove the clutch assembly from the base plate.

Refacing the Driven Plate

Should the facings of the driven plate require renewal, each rivet should be removed by using a $\frac{5}{32}$ in. diameter drill. The rivets should not be punched out.

Rivet one new facing in position, using a blunt ended centre punch, if the correct tool is not available, to roll the rivet shanks securely against the plate.

The second facing should then be riveted on the

opposite side of the plate with the clearance holes over the heads already formed in fitting the first facing.

The plate should then be mounted on a mandrel between centres and checked for "run out" as near the edge as possible; if the error is more than .015 in. press over at the high spots until it is true within this figure.

It is important to keep friction facings from oil or grease.

Refitting the Clutch

Place the driven plate on to the flywheel with the longer chamfered splined end of the driven plate hub towards the gearbox. The driven plate should be centralised by a dummy first motion shaft (Tool No. 18G 39) which fits the splined bore of the driven plate hub and the pilot bearing of the flywheel.

The clutch cover assembly can now be secured to the flywheel by means of the holding screws, tightening them a turn at a time by diagonal selection. There are two dowels in the flywheel to locate in the clutch cover. Remove the dummy shaft after these screws are fully tightened.

Finally remove the dummy shaft and refit the withdrawal bearing and gearbox. The weight of the gearbox must be supported during refittiing in order to avoid strain on the first motion shaft and distortion of the driven plate assembly.

Finally adjust the clutch pedal for free movement.

SERVICE DIAGNOSIS GUIDE

Symptom	No.	Possible Cause
(a) Drag or Spin	1 2 3 4 5 6 7 8 9 10	Oil or grease on driven plate linings Bent engine backplate leading to driven plate failure Misalignment between engine and first motion shaft Incorrect pedal adjustment or travel Driven plate hub binding on first motion shaft splines First motion shaft binding on its spigot bush Distorted clutch plate Warped or damaged pressure plate or clutch cover Broken clutch plate linings Dirt or foreign matter in clutch
(b) Fierceness or Snatch	 1 2	Check 1, 2 and 3 in (a) Binding clutch pedal mechanism Worn clutch linings
(c) Slip	 1 2	Check 1, 2 and 3 in (a) Check 1 in (b) Weak thrust springs Weak anti-rattle springs
(d) Judder	 1 2 3 4 5 6 7 8 9	Check 1, 2 and 3 in (a) Pressure plate out of parallel with flywheel face Friction facing contact area not evenly distributed Bent first motion shaft Buckled driven plate Faulty engine or gearbox rubber mountings Worn shackles Weak rear springs Propeller shaft bolts loose Loose rear spring clips
(e) Rattle	 1 2 3 4 5	Check 3 in (d) Damaged driven plate, i.e., broken springs, etc. Worn parts of release mechanism Excessive transmission backlash Wear in transmission bearings Release bearing loose on fork
(f) Tick or Knock	1 2 3 4 5	Worn first motion shaft bush Badly worn centre plate hub splines Out of line thrust plate Faulty bendix drive on starter Loose flywheel
(g) Driven Plate Fracture	 1	Check 2 and 3 in (a) Drag and metal fatigue due to hanging gearbox in driven plate

GEARBOX

GENERAL DATA

Type	Synchromesh on 1st, 2nd, top
Gear Control	Direct Gear Lever
Number of Gears	3 forward, 1 reverse
Type of Gears	Helecal Constant Mesh
Oil Capacity (including overdrive)	5½ pints; 6.6 U.S. pints, 3.12 litres
Gear Ratios :	
1st speed	2.25 : 1
2nd speed	1.42 : 1
3rd speed	Direct
Reverse	4.98 : 1
Overall Gear Ratios :	
1st speed	9.28
2nd speed	5.85
2nd Speed and Overdrive (32.4% reduction)	4.43
2nd Speed and Overdrive (28.6% reduction)	4.56
3rd Speed	4.125
3rd Speed and Overdrive (32.4% reduction)	3.12
3rd Speed and Overdrive (28.6% reduction)	3.28
Reverse	20.6

BEARINGS

Layshaft, Front :	
Type	Clevite
Length	1.75–1.76 in. (4.445–4.470 cm.)
Diameter (inside free)	0.868–0.874 in. (2.21.4–2.219 cm.)
Diameter (outside unfitted)	1.049 in. (2.664 cm.)
Diameter to be fitted into	1.046–1.047 in. (2.656–2.659 cm.)
Layshaft, Rear :	
Type	Clevite
Length	1.395–1.405 in. (3.543–3.568 cm.)
Diameter (inside free)	0.678–0.685 in. (1.722–1.739cm.)
Diameter (outside unfitted)	1.815 in. (4.610cm.)
Diameter to be fitted into	0.812–0.813 in. (2.062–2.065cm.)
Third Motion Shaft, Centre :	
Make	R & M
Type	M.J.1¼
Size	1¼×3⅛×⅞ in. (31.75×79.375×22.225 mm.)
First Motion Shaft :	
Make	R & M
Type	M.J.35G–three dot
Size	35×80×21 mm.

REMOVING THE GEARBOX AND
OVERDRIVE COMPLETE

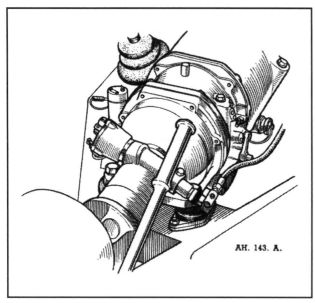

*Fig. 1.   Rear end of the gearbox and overdrive showing the rubber
mountings and speedometer flexible cable attachment.*

Before commencing removal operations the car
should be placed over a servicing pit.

First disconnect the battery by switching off the
master switch located at the rear left-hand side within
the boot.   Disconnect the starter lead at the starter.

Unscrew the gear lever knob and remove with locknut.   For ease of working remove the seat cushions.
Then unclip the padded central armrest and remove the
short piece of propeller shaft tunnel followed by the
gearbox cover and the bridge piece bulkhead cover.
Details for removing the latter are given on page P/8.
The rubber cover located at the bottom of the gear lever
should come away with the gearbox cover as it is
threaded over the lever.

Access is thus gained to the three nuts threaded on
studs mounted on the gear lever cup; with the removal
of these nuts the cup may be withdrawn together with
the three washers (2 rubber, 1 metal) and three distance
pieces located on the studs.

Note:—These studs are of U.N.F. thread, the
remainder of the gearbox studs have B.S.F. threads.

The gear lever can now be withdrawn from the
gearbox.

Disengage the propeller shaft from the driving
flange by releasing the four securing nuts, bolts and
washers.   In order to turn the propeller shaft, to work
on the nuts, the car may be pushed backwards or forwards quite easily, providing the handbrake is off.

While working within the car disconnect the wires
at their terminals on the overdrive switches.   There is a
main overdrive switch mounted on the right of the gearbox on which there are two wires.   The overdrive solenoid switch is located on the left of the overdrive housing
and only one wire has its terminal there.   A centrifugal
switch, having two wires to be disconnected, is found
on the centre of the rear housing.   Release the union cap
of the speedometer cable, at the rear housing end, and
then extract the cable from the housing.

The two rear mounting points located on either side
of the rear housing may now be released.   Unscrew the
two large setpins placed centrally to the brackets and not
the four smaller setpins securing the brackets to the
chassis.

Next, remove the various bolts and setpins securing
the top of the clutch housing to the engine rear mounting plate.   This includes the top bolt securing the starter
to the gearbox.

It is now required to lift the rear end of the engine
in order to withdraw the gearbox satisfactorily.   A
purchase block will be found suitable if a hook is secured
to the rear end of the rocker cover.   If a purchase block
is not available a suitable jack or the necessary packing
will give the required support.   This action also relieves
the load on the engine front mounting brackets when
the gearbox and overdrive becomes free.

The following dismantling is carried out under the
gearbox.

*Fig. 2.   The underside of the gearbox showing 1. Stabilising bar.
2. Gearbox cross shaft switch for overdrive. 3. Clutch operating shaft
lubricator. 4. Drain plug. 5. Overdrive drain plug.*

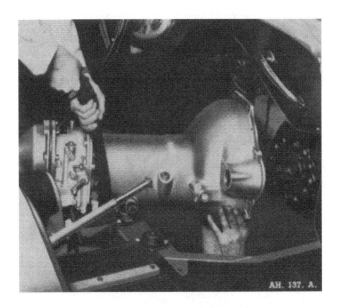

Fig. 3. Lifting the gearbox upwards and clear of the car.

Release the stabilising bar under the gearbox, and disconnect the clutch shaft from its linkage by removing the clutch shaft operating lever away from the shaft. The lever is secured by a nut and washer, and will be found on the right-hand side of the gearbox for right-hand drive models, and on the left-hand side for left-hand drive models.

Unscrew the remaining bolts and setpins holding the bottom of the clutch housing to the engine rear mounting plate, remembering that one of the bolts secures the starter. The latter can now be lifted out of the way.

Working inside the car again the gearbox first motion shaft should be withdrawn from the flywheel bearing and clutch by gently easing rearwards the gearbox and overdrive complete. If the unit does not detach itself readily it is probably because the rear of the engine requires raising still further.

Replacing the gearbox and overdrive is a reversal of the removal procedure.

DISMANTLING THE GEARBOX

Although the gearbox has only three operative speeds, it will be treated as a four speed box for dismantling procedure as the unused first gear is still in place. In the following narrative the latter gear is referred to as a spur gear.

Drain Plug

First drain the oil from the gearbox by removing the drain plug. The latter is situated beneath the gearbox at the left-hand side.

Gear Lever Control Box

With the gear lever already removed (see Gearbox Removal), the six bolts and one nut can be unscrewed which secure the control box in place and the control box taken off the main gearbox casing.

Cross Shaft Socket Lever

With the control box taken off access is gained to the cross shaft socket lever. Securing the lever to the shaft is a cotter pin, spring washer and nut; remove nut and washer, tap out the pin, then take the lever from the shaft.

Side Cover and Change Speed Gate

Holding the side cover in place are three setpins. The change speed gate is located by its two rounded ends, of the outer face, fitting into recesses in the gearbox side face. To release the gate from position it merely needs a gentle prising outwards with the aid of a screwdriver.

Selector Gate and Cross Shaft

The selector gate is secured to its spindle by a nut, which in turn is locked by a tab washer. Bend back the tab of the washer and release the securing nut. Periodically, to assist the latter operation, it is necessary to withdraw the selector gate spindle as one works at the nut, thus giving the nut removal clearance. The spindle oil seal and felt washer may be left in the housing.

Lift out the selector gate and then withdraw the cross shaft from its position complete with gear engagement lever.

It would be advisable at this stage to unscrew the overdrive control switch with its operating plunger on the opposite side of the gearbox.

Clutch Rod and Fork

Remove the nut and washer from the end of the operating shaft and lift off the operating arm. From within the clutch housing, release the nuts and spring washers from the clutch fork cotter pins, then tap the cotters from the operating fork. The clutch shaft may now be withdrawn from the housing, there being no need to disengage the circlips and washer from its left-hand side. If the car being operated on has left-hand drive, then the clutch shaft circlip and washer will be situated on the right-hand end of the shaft.

Front Cover

Release the front cover situated within the clutch housing by removing its seven ₁₆in. nuts and spring washers. It is not advisable at this stage of dismantling to endeavour to remove the cover with its joint washer, as the operation will prove far easier when the shaft fork selector rods are tapped forward, thus pushing the front cover away from the casing.

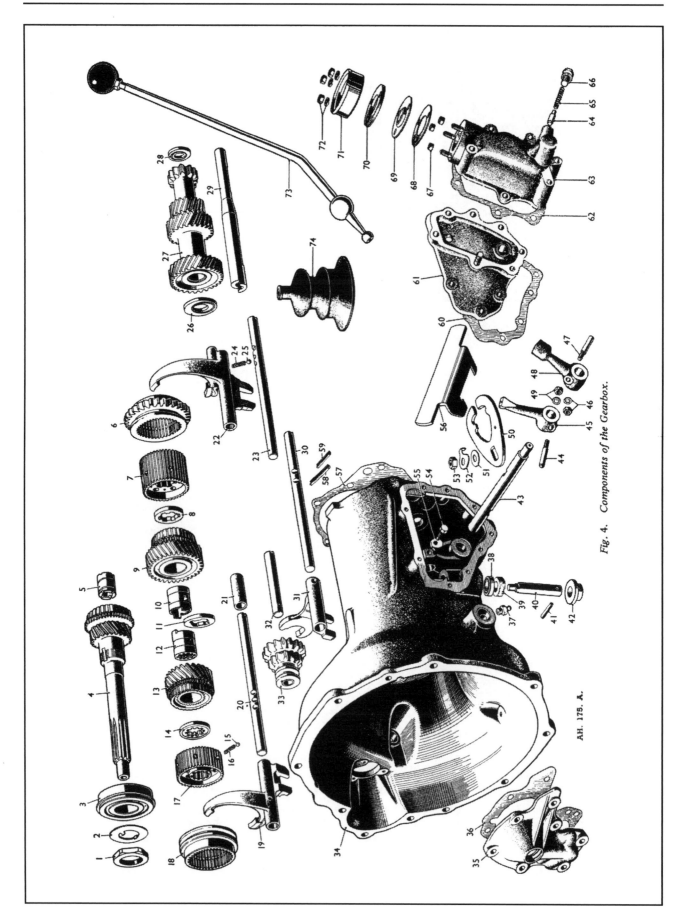

Fig. 4. Components of the Gearbox.

AH. 175. A.

Caption for Fig. 4.

1. *First motion shaft nut.*	26. *Thrust washer.*	51. *Washer.*
2. *Lockwasher.*	27. *Laygears.*	52. *Lockwasher.*
3. *First motion shaft bearing.*	28. *Thrust washer.*	53. *Spindle nut.*
4. *First motion shaft.*	29. *Lay shaft.*	54. *Reverse shaft pin.*
5. *Bush bearing.*	30. *Reverse selector rod.*	55. *Lockwasher.*
6. *Spur gear.*	31. *Reverse fork.*	56. *Change speed gate.*
7. *Second synchroniser.*	32. *Reverse gear shaft.*	57. *Rear joint washer.*
8. *Thrust washer.*	33. *Reverse gear.*	58. }*Selector rod keys.*
9. *First gear.*	34. *Gearbox casing.*	59. }
10. *Coupling sleeve.*	35. *Front cover joint washer.*	60. *Side cover joint.*
11. *Thrust washer.*	36. *Front cover.*	61. *Side cover.*
12. *Coupling sleeve.*	37. *Lubricator.*	62. *Control box joint.*
13. *Second gear.*	38. *Oil seal.*	63. *Control box.*
14. *Locking washer.*	39. *Felt ring.*	64. *Reverse stop plunger.*
15. *Ball.*	40. *Selector spindle.*	65. *Spring.*
16. *Spring.*	41. *Spindle pin.*	66. *Plug.*
17. *Second and 3rd synchroniser.*	42. *Spindle collar.*	67. *Distance pieces.*
18. *Synchroniser sleeve.*	43. *Cross shaft.*	68. *Rubber joint washer.*
19. *Second and 3rd fork.*	44. *Cotter pin.*	69. *Steel joint washer.*
20. *Second and 3rd selector rod.*	45. *Gear engagement lever.*	70. *Rubber washer.*
21. *Distance piece.*	46. *Cotter nut and washer.*	71. *Cup.*
22. *First speed fork.*	47. *Cotter pin.*	72. *Nuts and washers.*
23. *First selector fork.*	48. *Gear lever socket.*	73. *Gear lever.*
24. *Spring.*	49. *Cotter nut and washer.*	74. *Rubber dust cover.*
25. *Ball.*	50. *Selector gate.*	

Adapter Plate

Once the overdrive unit has been separated from the gearbox (see page G/8), the removal of the adapter plate is accomplished by first unscrewing the three nuts and washers mounted in recesses at the rear of the adapter plate and then the five setpins and tab washers on the outside face itself.

The overdrive pump cam should slide freely along the third motion shaft thus giving access to the circlip holding the distance piece to the rear adapter plate. Remove the circlip and slide the distance piece off the shaft. The adapter plate should now pull away from the gearbox, together with the rear main bearing. It may be necessary for one operator to hold the gearbox vertically by the adapter plate whilst a second operator taps the third motion shaft until the ball race in the adapter plate is free of the shaft.

Selector Rods and Forks

Working from the rear of the gearbox, use a soft metal drift, tap forward, for a short distance, each of the three selector rods and prise out the keys which are fitted to prevent the rods from turning.

This operation of tapping the rods forward should have eased the front cover from its place, making its removal, complete with joint washer, much easier.

Now drive each selector rod forward clear of the forks and extract them from the front of the gearbox. Care should be exercised in order that the spring loaded locating ball of each fork is not lost during this operation.

Lift out the three bronze forks, noting carefully their respective positions so that the assembly will be made more easy. Fitted behind the second speed fork

is a distance piece, which must be retrieved from the box when removing the respective fork.

Reverse Gear

The next dismantling operation is to remove the reverse gear. A lug which is an integral part of the main casting, locates the forward end of the reverse gear shaft. To secure the shaft in position, a setpin is screwed through the lug to locate in the shaft. The setpin is locked by a tab washer. Straighten the tab washer, release the setpin, then tap forward and remove the reverse gear shaft. Lift out the reverse gear.

Layshaft and Laygears

Working from the rear of the gearbox, use a bronze, or other soft metal drift, and drive the layshaft forward and out of the gearbox. The laygear cluster and the two thrust washers should drop to the bottom of the gearbox. These gears can only be lifted from the gearbox when the third and first motion shafts, together with their respective gears, have been removed.

Third Motion Shaft

The third motion shaft can now be withdrawn from the gearbox casing. The shaft should be eased rearwards, keeping it centralised, until the shaft spigot has cleared the bearing in the end of the first motion shaft, and then completely remove from position.

To remove the gears from the third motion shaft, first slide off the third and top speed (actual 2nd and top) synchronizer assembly. Then depress the spring loaded steel plunger, which locates the splined washer at the forward end of the third motion shaft, and turn the washer into line with the splines of the shaft. The third and second speed (actual 2nd and 1st) constant mesh gears, together with their common phosphor bronze sleeves and intervening thrust washer, can now be pulled over the steel plunger and so clear of the third motion shaft. Remove the steel plunger and its spring from the shaft.

Next remove the splined washer separating the first speed constant mesh gear assembly (not now used for forward speed) from the end spur gear unit, and then slide the spur gear assembly free from the third motion shaft.

If it is desired to dismantle the top and second (actual) speed coupling sleeve, or the spur gear assembly, these can be pressed clear of their splined synchronisers, but care must be taken to retrieve the balls and springs in each assembly.

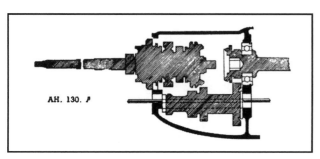

Fig. 5. *Withdrawing the third motion shaft through the rear opening of the gearbox. As an alternative to dropping the laygears to the base of the box they may be supported by a thin rod.*

Note:—Top and second synchro has three locating balls and springs and the first has six.

The bearing may be removed from the adapter plate by first taking out the circlip and then driving the bearing forward with a soft metal drift.

Lift out the third motion shaft front bearing from the first motion shaft.

First Motion Shaft

Before driving the first motion shaft from its position, tilt the laygears, now in the bottom of the gearbox, to clear the first motion shaft gear. Using a soft drift, inserted through the third motion shaft opening, drive the first motion shaft forward, complete with bearing and circlip, from the gearbox.

The laygears may now be removed from the gearbox.

To remove the bearing from the first motion shaft, knock back the tab washer and unscrew the shaft nut. This nut has a left-hand thread. The bearing may now be driven from the shaft, preferably by resting the circlip of the outer face on the jaws of an open vice and driving the shaft downwards. Use a hide or lead hammer for the operation, as great care must be exercised to prevent the end of the first motion shaft from "spreading" due to the hammering.

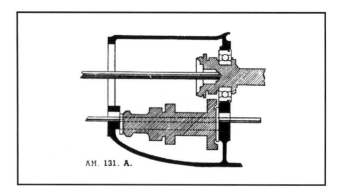

Fig. 6. *Driving out the first motion shaft with a drift acting through the housing for the third motion shaft front bush bearing. Take care not to damage this housing.*

EXAMINATION FOR WEAR

Clutch Cross Shaft Bushes

It is unlikely that these bushes should have worn, but should the cross shaft appear excessively loose in the bushes, new ones must be fitted. These bushes are fitted in two parts with a gap left between the parts, thus providing an oil channel for lubrication via the oiling nipples.

Bearings

The first and third motion shaft ball bearings may become worn after a considerable length of service and should be renewed if there are signs of looseness between inner and outer races.

First and Third Motion Shaft Bush

This bush is fitted with a maximum permissible internal clearance of .003 in. for the third motion shaft. When any appreciable wear above this figure occurs, the bush and shaft should be examined and renewed where necessary.

Third Motion Shaft Sleeve

The phosphor bronze sleeve which carries the first and second (actual) speed gear assemblies must be replaced if the wear between the shaft end sleeve appears excessive. The fitted clearance in a new gearbox is between .00025 and .00175 in.

Laygear Thrust Washers

These washers are designed to permit an end float for the layshaft cluster gears between .001 and .003 in. If the end float exceeds this tolerance, the thrust washer must be renewed. The smaller thrust washer, positioned at the rear, is made in varying thicknesses to allow for correct end float to be obtained.

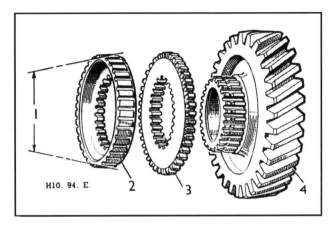

Fig. 7. Synchronising cone.
1. Chamfer to be machined after cone is shrunk into position. 2. Cone
3. Coupling adapter. 4. Constant mesh gear.

Layshaft and Bushes

The layshaft and layshaft bushes, in the cluster gear assembly, may become worn and need renewal. Both the front and rear bushes have an internal clearance of .002 in.–.003 in. These bushes are a press fit in the laygear.

Gear Synchronising Cones

These cones are "shrunk on" to the first, second and third speed gears (actual), which are normally supplied as a complete unit for spares purposes. Where facilities exist for shrinking on and final machining, cones can be supplied separately. However, care must be taken in fitting if the gear is to operate satisfactorily.

The internal broaching of the cone is calculated to allow for a shrinkage fit on to the gear serrations, and the cone must be heat-expanded before it can be fitted.

When heated to approximately 250 degrees Fahrenheit, expansion will allow the cone to be pressed home on to the gear without damaging the broaching and will

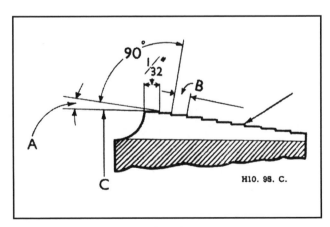

Fig. 8. Cone dimensions.
A. Cone angle 5° 10′. B. Coarse thread turned with ·015 in. lead.
C. Cone diameter 2·5 in.

be sufficiently close fitting to resist displacement in gear changing.

The heating can best be done by immersion in oil of 250 degrees Fahrenheit and then fitting by means of a hand press. After shrinking on, the unit should be immediately quenched in water to prevent the heat softening the gear itself.

On each gear the appropriate speed coupling adapter must be fitted before the cone, but there is no need to pre-heat this adaptor, which can be pressed home in the cold state. There is a shoulder on one side of the adapter, and this must be facing the gear and not the cone.

When the cone is in position, the final machining can be done in accordance with the dimensions given in fig. 8. The taper of the cone must be true and concentric with the bore to .001 in.

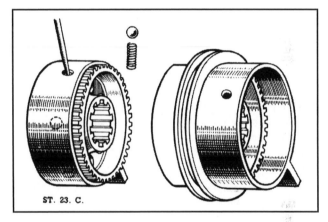

Fig. 9. Using Service Tool 18G 40 to assemble the spring-loaded balls to a coupling sleeve and synchroniser.

ASSEMBLING THE GEARBOX

Perfect cleanliness of the gearbox parts is essential before commencing assembly. Although the following complete assembly operation has been sub-divided, it is advisable for the operator to read the whole description before commencing any work. The sub-assemblies of the various parts are so interlaced with one another.

To reassemble the gearbox proceed as follows :—

Synchromesh Sub-Assembly

During manufacture both the spur gear and the 2nd and 3rd (actual) speed coupling sleeves are each paired with their respective synchronisers. Only mated pairs of these parts should therefore be fitted.

Special guides are available to facilitate the reassembling of the three balls and springs into the synchronisers. The guide is of the same diameter as the coupling sleeve, see fig. 9.

The guide is slipped over the synchroniser and turned until the hole coincides with one of the ball sockets. A spring and ball are then placed in position, the ball depressed and the guide rotated to hold it in place. This procedure is repeated for each spring and ball in turn until they are all depressed. The guide is then pushed further along the synchroniser splines, followed by the coupling sleeve.

As the coupling sleeve replaces the guide, the balls find their correct location in the coupling sleeve groove.

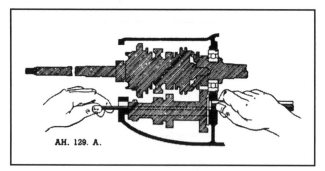

Fig. 10. Lifting the laygears into position and inserting the layshaft as the thin rod is withdrawn.

Layshaft Gears

First locate the two thrust washers to the laygears, ensuring that the larger washer is at the front, and then place the gear cluster in the gearbox. Check that there is an end play for the cluster gears of between .001 and .003 in., and remedy if necessary by fitting a thicker or thinner rear washer. Temporarily replace the layshaft with a thin rod which will permit the gear cluster to keep out of the mesh with the third and first motion shaft gears.

First Motion Shaft Gears

Press the ball bearing on to the first motion shaft with the circlip on the outer race of the bearing facing forwards. The bearing must be pressed onto the shaft as far as it will go. Refit the keyed washer and screw down and tighten the left-hand thread locking nut. Secure the nut with the locking washer.

Using a bronze drift, tap the bearing, complete with the first motion shaft, into the forward face of the gearbox casing until the bearing circlip is flush with its recess in the casing.

Third Motion Shaft

Lightly oil the large centre splines of the shaft and refit the spur gear assembly with the synchroniser pointing forward. Refit the keyed thrust washer on the shaft and assemble the first and second speed gears on to their phosphor bronze sleeve, which must be lightly oiled. These two sleeves are made solid by a common thrust washer (see fig. 4).

The second speed (actual), or small gear, must be placed on that end of the sleeve which has internal splines. Slide the sleeve and gears on to the third motion shaft with the second gear to the front.

Place the spring and the plunger into the hole in the third motion shaft and slide on the splined washer. Depress the plunger and slide the splined washer over the plunger, then turn the washer for the plunger to engage with one of the grooves in the washer.

The gears are now assembled on the third motion shaft.

Place the second and top (actual) synchroniser and coupling sleeve on to the third motion shaft, with the coupling sleeve groove for the change speed fork to the rear, and then oil and fit the phosphor bronze bush into the end of the first motion shaft.

Place the 2nd and 3rd synchronisers slightly forward on the shaft to clear the laygears, and then guide the third motion shaft assembly into the gearbox casing. When the front end of the shaft is firmly fixed into the allotted phosphor bronze bush in the first motion shaft, the layshaft gear cluster should be raised into mesh with the gears, and the layshaft oiled and fitted into position. The lipped end of the layshaft must face forwards, and the rear end must be flush with the gearbox casing.

Reverse Gear

Refit the reverse gear into the gearbox casing with the large gear to the rear. Oil the reverse gear shaft before inserting, and secure the shaft with the locating pin and tab washer.

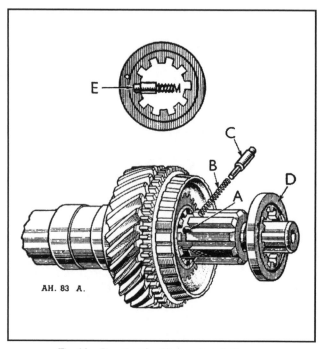

Fig. 11. Securing the third motion shaft gears.
A. Shaft hole for locking peg. B. Spring. C. Locking peg. D. Locking washer. E. Locking washer with peg engaged.

Selector Rods and Forks

Before commencing to locate the selector forks within the gearbox casing it is advisable to pre-load the spring and ball into each bronze fork and with the aid of a pilot bar, as in fig. 12, retain the spring and ball in position until each fork rod has entered its correlative fork.

With the gears in the neutral position, first fit the 2nd and top speed selector fork, and then locate the 1st speed selector. Now tap the 2nd and 3rd selector fork rod through the casing, slide the distance piece over the rod, and continue tapping the rod through its fork until it reaches its final position. Next locate the reverse gear fork and then enter the 1st selector fork rod and the reverse gear fork rod, through the casing and into their respective forks.

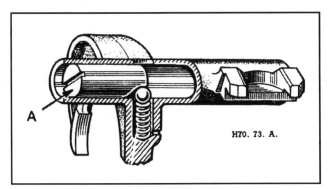

Fig. 12. Preloading a change speed fork ball and spring with the aid of a pilot bar 'A'.

Do not drive the two latter rods completely home until the change speed gate lever and its arm have been fitted to the box.

When driving the fork rods home remember to retrieve the pilot bars as they leave the forks.

Selector Gate and Cross Shaft

Slide the cross shaft, together with its engagement lever, into the allotted hole in the casing. Replace the selector gate spindle and fit the selector gate over the top of the spindle, locating it on the two flats. Make sure that the cross shaft engagement lever is between the arms for the selector gate and slides freely between the forks. Secure the selector gate by means of the nut and tab washer.

The selector fork rods may now be tapped right through the forks into their final position.

At this stage of the reassembly it may be permitted to screw in the overdrive switch and plunger into the opposite side of the casing.

Front Cover

The front cover with its paper joint washer should now be positioned over the securing studs and fixed with the seven $\frac{5}{16}$ in. nuts and spring washers.

At this stage of the reassembly, the selector fork rods should be locked in place with the two keys in the gearbox rear face.

Adapter Plate

Slide the adapter plate, together with its bearing and paper joint washer, along the third motion shaft locating the plate on three studs at the rear of the gearbox casing. Loosely fit the five setpins and tab washers and then the three nuts and spring washers. Do not tighten the five main setpins before the three nuts are fitted as the space available to place the nuts on their studs is restricted.

Tighten down all setpins and nuts and knock over the five tab washers.

Fit the distance piece which covers the space between the rear main bearing and the groove allocated for the circlip, and fix on the latter.

Note:—The overdrive pump cam is positioned when assembling overdrive to gearbox, see page G/11.

Change Speed Gate and Side Cover

The change speed gate should now be fitted into position. The gate is in the form of an angle plate, the side incorporating the gear stops slides into the gearbox (see fig. 4) and the rounded ends of the outer side of the gate locate in two recesses which are bored into the gearbox side face.

Next secure the side cover into position by means of the three setpins and spring washers, ensuring that the paper joint washer is undamaged.

Cross Shaft Socket Lever

Slide the cross shaft socket lever along the cross shaft with the crank of the former facing outwards. Secure it to the cross shaft with its cotter pin, spring washer and nut.

Control Box and Gear Lever Assembly

Place the control box with its joint paper washer onto the side cover locating it with a stud fitting in its respective hole, and screw in the six bolts and one nut together with spring washers. Observe, however, that there is one bolt shorter than the other five, this fitting into the hole located at the centre rear of the control box.

The gear lever should now be pushed down into the control box locating it with the spline on the ball of the lever and the locating pin in the control box itself. See that a good fit is made with the cross shaft socket lever.

Three small distance pieces which were removed from the top U.N.F. studs of the control box should now be replaced together with one metal and two rubber washers—see fig. 4 for order of assembly. The gear lever cup can now take its place over the washers and be tightened down by means of the three U.N.F. nuts and spring washers. Screw in the spring and plunger located at the bottom rear of the control box.

Note:—The rubber cover should be replaced with the gearbox cover, then refit the knob and locknut.

Clutch Shaft and Fork

Slide the clutch shaft through the rear left-hand side of the clutch housing (right-hand side for left-hand models), then position the clutch operating yoke on the shaft and pass the shaft through the right-hand side of the casing. Secure the yoke in place with the two cotter pins, washers and nuts, and fix the operating lever to the end of the shaft by means of the washer and nut.

SERVICE DIAGNOSIS GUIDE

Symptom	No.	Possible Cause
(a) Jumping out of Gear	1	Broken change speed fork rod spring
	2	Excessively worn fork rod groove
	3	Worn coupling dogs
	4	Fork rod securing screw loose
(b) Noisy Gearbox	1	Insufficient oil in gearbox
	2	Excessive end play in laygear
	3	Damaged or worn bearings
	4	Damaged or worn teeth
(c) Difficulty in Engaging Gear	1	Incorrect clutch pedal adjustment
(d) Oil Leaks	1	Damaged joint washers
	2	Damaged or worn oil seals
	3	Front, rear or side covers loose or damaged

SERVICE JOURNAL REFERENCE

NUMBER	DATE	SUBJECT	CHANGES

OVERDRIVE

GENERAL DATA

Component	Dimensions New	Clearance New
Pump		
Plunger Diameter	$\frac{3}{8}$ in.−.0004 (9.525 mm.−.0102) −.0008 (−.0203)	
Bore for Plunger...	$\frac{3}{8}$ in.+.0008 (9.525 mm.+.0203) −.0000 (−.0000)	+.0016 in. (.0406 mm.) +.0004 in. (.0102 mm.)
Plunger spring (fitted load at top of stroke)	9 lb. 12$\frac{3}{4}$ oz. (4.444 kg.)	
Valve spring load	5 lb. at $\frac{3}{16}$ in. long (2.268 kg. at 4.763 mm.)	
Pin for roller	$\frac{1}{4}$ in.±.00025 (6.35 mm.±.00635)	
Bore for pin in roller	$\frac{1}{4}$ in.+.002 (6.35 mm. +.0508) −.001 (−.0254)	+.00225 in. (.0572 mm.) +.00075 in. (.0191 mm.)
Gearbox Third Motion Shaft		
Shaft diameter at steady bushes ...	1$\frac{5}{32}$ in.−.0009 (29.369 mm.−.0229) −.0018 (−.0457)	
Steady bush internal diameter... ...	1$\frac{5}{32}$ in.+.003 (29.369 mm.+.0762) +.002 (+.0508)	+.0048 in. (.1219 mm.) +.0011 in. (.0279 mm.)
Shaft diameter at sun wheel bush ...	1$\frac{5}{32}$ in.−.0009 (29.369 mm.−.0229) −.0018 (−.0457)	
Sun wheel bush internal	1$\frac{5}{32}$ in.+.0012 (29.369 mm.+.0305) −.0000 (−.0000)	+.0030 in. (.0762 mm.) +.0009 in. (.0229 mm.)
Shaft diameter at rear steady bush ...	$\frac{5}{8}$ in.−.0008 (15.875 mm.−.0203) −.0015 (−.0381)	
Rear steady bush internal diameter ...	$\frac{5}{8}$ in.+.001 (15.875 mm.+.0254) −.000 (−.0000)	+.0025 in. (.0635 mm.) +.0008 in. (.0203 mm.)
Gear Train		
Planet pinion bush internal diameter...	$\frac{3}{4}$ in.+.001 (19.05 mm.+.0254) −.0000 (−.0000)	
Planet bearing shaft external diameter...	$\frac{3}{4}$ in.−.0015 (19.05 mm. −.0387) −.0020 (−.0508)	+.003 in. (.0762 mm.) +.0015 in. (.0381 mm.)
End float of sun wheel		.008 to .014 in. (.2032 to .3556 mm.)
Piston Bores		
Accumulator bores	1$\frac{1}{8}$ in.±.0005 (28.575 mm.±.0127)	
Operating piston bore	1$\frac{1}{8}$ in.±.0005 (28.575 mm.±.0127)	
Clutch		
Movement from direct to overdrive ...	−$\frac{1}{16}$ in. (1.588 mm.)	
Allowance for wear direct drive clutch...	−$\frac{1}{8}$ in. (3.175 mm.)	
Allowance for wear overdrive clutch ...	−$\frac{1}{8}$ in. (3.175 mm.)	

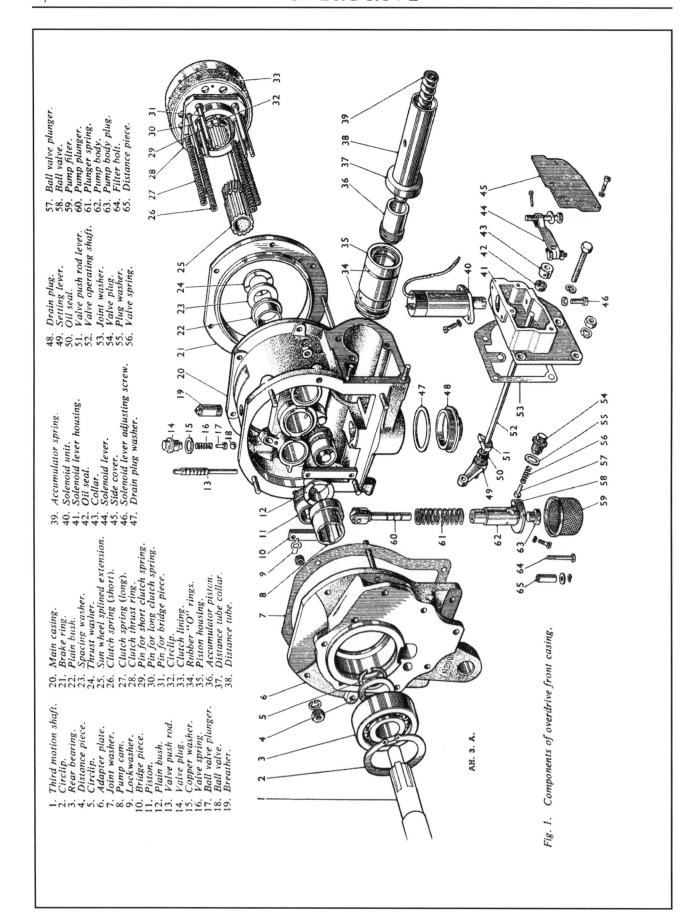

1. Third motion shaft.
2. Circlip.
3. Rear bearing.
4. Distance piece.
5. Circlip.
6. Adapter plate.
7. Joint washer.
8. Pump cam.
9. Lockwasher.
10. Bridge piece.
11. Piston.
12. Plain bush.
13. Valve push rod.
14. Valve plug.
15. Copper washer.
16. Valve spring.
17. Ball valve plunger.
18. Ball valve.
19. Breather.

20. Main casing.
21. Brake ring.
22. Plain bush.
23. Spacing washer.
24. Thrust washer.
25. Sun wheel splined extension.
26. Clutch spring (short).
27. Clutch spring (long).
28. Clutch thrust ring.
29. Pin for short clutch spring.
30. Pin for long clutch spring.
31. Pin for bridge piece.
32. Circlip.
33. Clutch lining.
34. Rubber "O" rings.
35. Piston housing.
36. Accumulator piston.
37. Distance tube collar.
38. Distance tube.

39. Accumulator spring.
40. Solenoid unit.
41. Solenoid lever housing.
42. Oil seal.
43. Collar.
44. Solenoid lever.
45. Side cover.
46. Solenoid lever adjusting screw.
47. Drain plug washer.

48. Drain plug.
49. Setting lever.
50. Oil seal.
51. Valve push rod lever.
52. Valve operating shaft.
53. Joint washer.
54. Valve plug.
55. Plug washer.
56. Valve spring.

57. Ball valve plunger.
58. Ball valve.
59. Pump filter.
60. Pump plunger.
61. Plunger spring.
62. Pump body.
63. Pump body plug.
64. Filter bolt.
65. Distance piece.

AH. 3. A.

Fig. 1. Components of overdrive front casing.

ROUTINE MAINTENANCE AND DESCRIPTION

LUBRICATION

The lubricating oil in the overdrive unit is common with that in the gearbox and the level should be checked with the gearbox dipstick.

It is essential that an approved lubricant be used when refilling. Trouble may be experienced if some types of extreme pressure lubricants are used because the planet gears act as a centrifuge to separate the additives from the oil.

Recommended lubricants are given in section S. It should be emphasised that any hydraulically controlled transmission must have clean oil at all times and great care must be taken to avoid the entry of dirt whenever any part of the casing is opened.

Every 1,000 miles (1,600 kilometres) check the oil level of the gearbox and overdrive and top up if necessary through the dipstick hole.

Every 6,000 miles (9,600 kilometres) drain and refill the gearbox and overdrive unit. In addition to the normal drain plug fitted to the gearbox the overdrive unit incorporates a plug at its base which gives access to a filter. This plug should also be withdrawn to ensure that all used oil is drained away from the system.

Every 6,000 miles (9,600 kilometres) after draining the oil, remove the overdrive oil pump filter and clean the filter gauze by washing in petrol. The filter is accessible through the drain plug hole and is secured by a central set bolt.

Refilling of the complete system (gearbox and overdrive) is accomplished through the gearbox filler plug. The capacity of the combined gearbox and overdrive unit is 5½ pints (6.6 U.S. pints; 3.12 litres).

After draining, ¼ pint of oil will remain in the overdrive hydraulic system, so that only 5¼ pints will be needed for refilling. If the overdrive has been dismantled the total of 5½ pints will be required.

After refilling the gearbox and overdrive with oil, recheck the level after the car has been run, as a certain amount of oil will be retained in the hydraulic system of the overdrive unit.

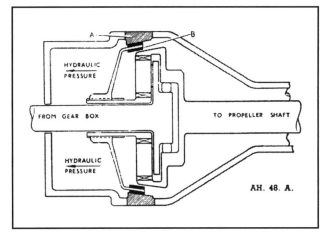

Fig. 3. Overdrive engaged.

WORKING DESCRIPTION

The overdrive unit comprises a hydraulically controlled epicyclic gear housed in a casing which is directly attached to an extension at the rear of the gearbox.

The synchromesh gearbox third motion shaft is extended and carries at its end the inner member of an uni-directional clutch (see Fig. 2). The outer member of this clutch is carried in the combined annulus and output shaft.

Also mounted on the third motion shaft are the planet carrier G and a freely rotatable sun wheel. Splined to a forward extension E of the sun wheel and sliding thereon is a cone clutch member D, the inner lining of which engages the outside of the annulus F while the outer lining engages a cast-iron brake ring sandwiched between the front and rear parts of the unit housing.

A number of compression springs is used to hold the cone clutch in contact with the annulus, locking the sun wheel to the latter so that the entire gear train rotates as a solid unit, giving direct drive. In this condition the drive is taken through the uni-directional clutch, the cone clutch taking over-run and reverse torque, as without it there would be a free-wheel condition.

The spring pressure can be overcome through the medium of two pistons, working in cylinders formed in the unit housing, supplied with oil under pressure from a hydraulic accumulator. This hydraulic pressure causes the cone clutch to engage the stationary brake ring (A Fig. 3) and bring the sun wheel to rest, allowing the annulus to over-run the uni-directional clutch and give an increased speed to the output shaft, i.e. "overdrive".

When changing from overdrive to direct gear, if the accelerator pedal is released (as in a change down for engine braking) the cone clutch, being oil immersed, takes up smoothly. If the accelerator pedal is not released, when contact between the cone clutch and brake

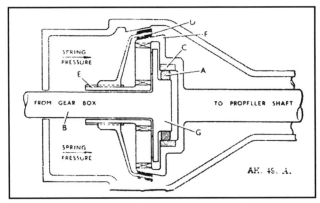

Fig. 2. Overdrive disengaged.

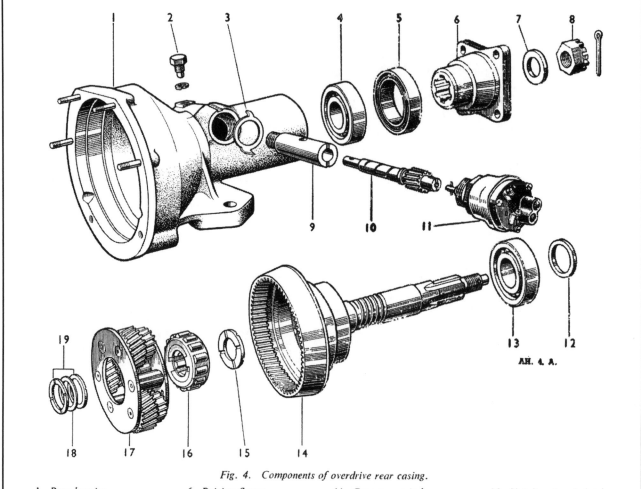

Fig. 4.   Components of overdrive rear casing.

1. *Rear housing.*	6. *Driving flange.*	11. *Governor switch.*	16. *Uni-directional clutch.*
2. *Locking peg for "9".*	7. *Washer.*	12. *Spacing washer.*	17. *Planet carrier.*
3. *Lockwasher.*	8. *Flange nut.*	13. *Inner bearing.*	18. *Spacing washer.*
4. *Outer bearing.*	9. *Speedometer pinion sleeve.*	14. *Annulus.*	19. *Thrust washer.*
5. *Oil seal.*	10. *Speedometer pinion.*	15. *Thrust washer.*	

ring is broken, the unit still operates momentarily in its overdrive ratio as engine speed and road speed remain unchanged. But the load on the engine is released and it begins to accelerate, speeding up the sun wheel from rest until, just at the instant when its speed synchronises with the speed of the annulus, the whole unit revolves solidly and the uni-directional clutch takes up the drive once more. The movement of the cone clutch is deliberately slowed down so that the uni-directional clutch is driving before the cone clutch contacts, ensuring a perfectly self-synchronised change.

CONSTRUCTION

The third motion shaft of the synchromesh gearbox is extended to carry first a cam operating the oil pump and then a steady bearing with two opposed plain bushes carried in the front housing. Next is the sun wheel of the epicyclic gear carried on a Clevite bush, and beyond this the shaft is splined to take the planet carrier and uni-directional clutch. The end of the shaft is reduced and carried in a plain bush in the output shaft. The latter is supported in the rear housing by two ball bearings. The clutch member slides on the splines of the sun wheel extension to contact either the annulus or a cast iron brake ring forming part of the unit housing.

To the hub of the cone clutch member is secured a ball bearing housed in a flanged ring. This ring carries on its forward face a number of pegs acting as guides to compression springs by which the ring, and with it the clutch member, is held against the annulus. The springs prevent free-wheeling on over-run and are of sufficient strength to handle reverse torque. Also secured to the ring are four studs picking up two bridge pieces against which bear two pistons operating in cylinders formed in the unit housing. The cylinders are connected through a valve to an accumulator in which pressure is maintained by the oil pump. The operating pistons are fitted with special three-piece cast-iron rings, as also is the accumulator piston.

When the valve is open, oil under pressure is admitted to the cylinders and pushes the pistons forward to engage the overdrive clutch. Closing the valve cuts off the supply of oil to the cylinders and allows it to escape. Under the influence of the springs the clutch member moves back to engage direct drive position. The escape of oil from the cylinders is deliberately restricted so that the clutch takes about half a second to move over.

The sun wheel and pinions are cyanide case-hardened and the annulus heat-treated. Gear teeth are helical. The pinions have Clevite bushes and run on case-hardened pins.

The outer ring of the uni-directional clutch is pressed and riveted into the annulus member. The clutch itself is of the caged roller type, loaded by a clock-type spring made of round wire.

The hydraulic system is supplied with oil by a plunger type pump operated by a cam on the gearbox third motion shaft. The pump body is pressed into the front housing and delivers oil through a non-return valve to the accumulator cylinder, in which a piston moves back against a compression spring until the required pressure is reached when relief holes are uncovered. From the relief holes the oil is led through drilled passages to an annular groove between the two steady bushes on the gearbox third motion shaft.

Radial holes in the shaft collect the oil and deliver it along an axial drilling to other radial holes in the shaft from which it is fed to the sun wheel bush, thrust washers, planet carrier and planet pins.

From the accumulator, oil under pressure is supplied to the operating valve chamber. This forms an enlargement at the top of a vertical bore and contains a ball valve, the ball seating downwards thus preventing oil from circulating to the operating cylinders. The valve is a hollow spindle sliding in the bore, its top end reduced and carrying a seating for the ball, which is then lifted, admitting oil to the operating cylinders and moving the pistons forward to engage the overdrive clutch.

When the valve is lowered the ball is allowed to come on to its seating in the housing, cutting off pressure to the cylinders.

Further movement of the valve brings it out of contact with the ball, allowing the oil from the cylinders to escape down the inside of the valve to discharge into the sump. The cone member then moves back under the influence of the clutch springs.

SERVICING IN POSITION

When the overdrive does not operate properly it is advisable first to check the level of oil and, if below the requisite level, top up with fresh oil and test the unit again before making any further investigations.

Before commencing any dismantling operations it is important that the hydraulic pressure is released from the system. Do this by operating the overdrive 10–12 times.

As the unit is fitted with a speed responsive control it will be found more convenient to carry out this operation by moving the valve setting lever manually.

GUIDE TO SERVICE DIAGNOSIS

Overdrive Does Not Engage

1. Insufficient oil in box.
2. Electric control not operating.
3. Leaking operating valve due to foreign matter on ball seat or broken valve spring.
4. Pump not working due to choked filter.
5. Pump not working due to broken pump spring.
6. Leaking pump non-return valve due to foreign matter on ball seat or broken valve spring.

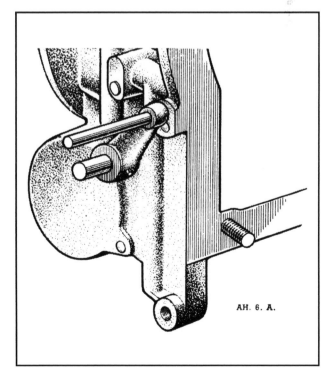

AH. 6. A.

Fig. 5. Valve setting lever.

7. Insufficient hydraulic pressure due to leaks or broken accumulator spring.
8. Damaged gears, bearings or moving parts within the unit requiring removal and inspection of the assembly.

Overdrive Does Not Release

1. Electric control not operating.
2. Blocked restrictor jet in valve.
3. Sticking clutch.
4. Damaged parts within the unit necessitating removal and inspection of the assembly.

Clutch Slip In Overdrive

1. Insufficient oil in gearbox.
2. Worn clutch lining.
3. Insufficient hydraulic pressure due to leaks.

Clutch Slip in Reverse or Free-Wheel Condition on Over-run

1. Worn clutch lining.
2. Blocked restrictor jet in valve.
3. Insufficient pressure on clutch due to broken clutch springs.

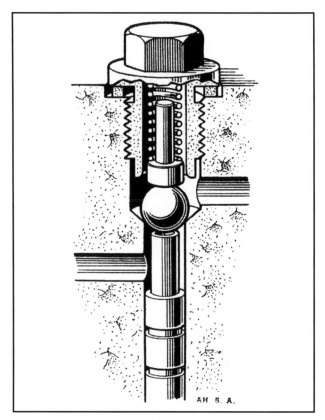

Fig. 6.   Operating valve.

OPERATING VALVE

Having gained access to the unit through the floor, unscrew the valve plug and remove the spring and plunger. The ball valve will then be seen inside the valve chamber. The ball should be lifted $\frac{1}{32}$ in. (.794 mm.) off its seat when the overdrive control is operated.

As the unit is fitted with a speed responsive control the appropriate parts of the electrical circuit must be shorted out in order to operate the control.

If the ball does not lift by this amount the fault lies in the control mechanism. Located on the right-hand side of the unit and pivoting on the valve operating cross shaft, which passes right through the housing, is a valve setting lever. In its outer end is a $\frac{3}{16}$ in. (4.763 mm.) diameter hole which corresponds with a similar hole in the housing when the unit is in "overdrive" (i.e. when the ball is lifted $\frac{1}{32}$ in. off the valve seat).

If the two holes do not line up, adjust the control mechanism until a $\frac{3}{16}$ in. diameter rod can be inserted through the setting lever into the hole in the housing. Check lift of ball after completing the adjustment.

A small magnet will be found useful for removing the ball from the valve chamber. The valve can be withdrawn by inserting the tang of a file into the top, but care must be taken not to damage the ball seating at the end of the valve. Near the bottom of the valve will be seen a small hole breaking through to the centre drilling. This is the jet for restricting the exhaust of oil

from the operating cylinders. Ensure that this jet is not choked.

HYDRAULIC SYSTEM

If the unit fails to operate and the ball valve is found to be seating and lifting correctly check that the pump is functioning.

Jack up the rear wheels of the car, then with the engine ticking over and the valve plug removed, engage top gear. Watch for oil being pumped into the valve chamber. If none appears then the pump is not functioning.

The pump (Fig. 7) described above, is of the plunger type and delivers oil via a non-return valve to the accumulator. Possible sources of trouble are (1) failure of the non-return valve due to foreign matter on the seat or to a broken valve spring and (2) breakage of the spring holding the pump plunger in contact with the cam.

The pump is self priming, but failure to deliver oil after the system has been drained and refilled indicates that the air bleed is choked causing air to be trapped inside the pump.

In the unlikely event of this happening it will be necessary to remove the pump and clean the flat on the pump body and the bore of the casting into which it fits.

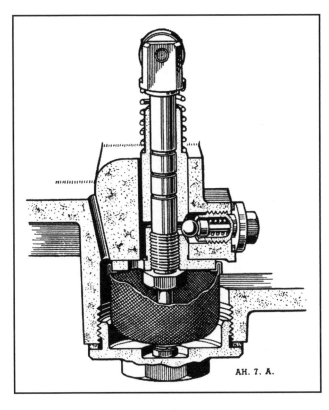

Fig. 7. The pump in cut-away form.

PUMP VALVE

Access to the pump valve is gained through a cover on the left-hand side of the unit. Proceed as follows:—

Solenoid Operated Units

1. Remove drain plug and drain off oil.
2. Remove cover from solenoid bracket.
3. Remove solenoid.
4. Slacken off clamping bolt in operating lever and remove lever, complete with solenoid plunger.
5. Remove distance collar from valve operating shaft.
6. The solenoid bracket is secured by two $\frac{5}{16}$ in. (7.938 mm.) studs and two $\frac{5}{16}$ in. diameter bolts, the heads of which are painted red, **remove the nuts from the studs before touching the bolts. This is important.** The two bolts should now be slackened off together, releasing the tension on the accumulator spring.
7. Remove the solenoid bracket.
8. Unscrew the valve cap and take out the spring, plunger and ball.

Reassembly is the reverse of the above operations. Ensure that the soft copper washer between the valve cap and pump housing is nipped up tightly to prevent oil leakage.

It will now be necessary to reset the valve operating lever. Proceed as follows:—

Before clamping up the valve operating lever and replacing the solenoid bracket cover, rotate the valve operating shaft until a $\frac{3}{16}$ in. (4.763 mm.) diameter pin can be inserted through the valve setting lever into the corresponding hole in the casing. Leave the pin in position, locking the unit in the "overdrive" position. Lift the solenoid plunger up to the full extent of its stroke (i.e. to its energised position) and clamp up the operating lever. Adjust the stop screw under the solenoid plunger until there is a clearance of $\frac{1}{4}$ in. (6.35 mm.) between the stop screw and the head of the solenoid plunger bolt (Fig. 8). This may be checked by using a piece of $\frac{1}{4}$ in. diameter rod as a gauge. Remove the pin through the setting lever and operate the lever manually to check that the control operates easily.

Check that when the solenoid is energised the hole in the valve setting lever corresponds with the hole in the casing. If it does not quite line up, a fine adjustment can be made by screwing the solenoid plunger bolt further in or out of the plunger. This will, of course, mean removal and replacement of the split pin between the two parts. After ensuring that the setting is correct replace the cover to the solenoid bracket.

To Dismantle the Pump

Proceed as follows:—

1. Remove the drain plug and drain off oil.
2. Remove pump valve as previously described.

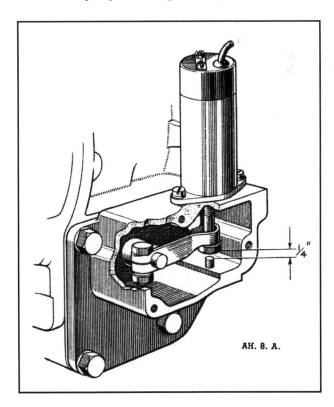

Fig. 8. Solenoid lever.
There should be a clearance of $\frac{1}{4}$" between bolt head and stop screw when solenoid plunger is raised.

AH. 164. A.

Fig. 9. Using extractor 18G 183 to remove the pump body.

3. Remove the filter after unscrewing the securing bolt.
4. Take out the two cheese head screws securing the pump body flange and extract the pump body. A special extractor tool (Fig. 9) is available for this purpose. This screws into the bottom of the pump body in place of the screwed plug.

Assembly of the Pump

Replace the plug in the bottom of the pump body, ensuring that it is screwed home tightly. Line up the pump body so that the inlet port and holes for securing screws register with the corresponding holes in the housing, and tap the pump body home.

The pump plunger is prevented from rotation when in position by a guide peg carried in the front casing. When assembling the pump the plunger should be inserted with the flat on its head facing the rear of the unit. It is possible to guide it past the guide peg by means of a screwdriver inserted through the side of the casting.

HYDRAULIC PRESSURE

A working oil pressure of 450–500 lbs. per sq. in. (31.635–31.15 kg./cm².) is required.

An adapter for use in connection with a pressure gauge is available from Messrs. Laycock Engineering Limited. This adapter should replace the plug which is screwed into the casing over the operating valve and should be used in conjunction with a pressure gauge reading up to a minimum of 800 lb. per sq. in. (56.24 kg./cm.²) and suitable for screwing into the $\frac{1}{8}$ in. B.S.F. thread in the mouth of the adapter.

Low pressure indicates leakage or possibly a broken accumulator spring.

To Remove the Accumulator Piston and Spring

1. Proceed as described in operations 1 to 8 in the paragraph headed "Pump Valve."
2. The spring and distance tube can now be withdrawn.
3. Using the special extractor tool 18G 182 remove the piston housing and piston.

It is important to appreciate that correct fitting of the piston rings is of vital importance to the efficient working of the unit. Check that the rings are not gummed up due to use of an unsuitable lubricant or have excessive clearance in the groove. Ensure that the rubber "O" rings on the piston housing are not damaged.

DISMANTLING AND REASSEMBLING THE UNIT

DISMANTLING

Should trouble arise necessitating dismantling of the unit to a degree further than has already been described, it will be necessary to remove the unit from the car.

Whilst it is possible to lift out the overdrive alone from the car, it is advised that the gearbox and overdrive be removed as a single unit. It is far easier to refit the overdrive to the gearbox when the assembly is on a bench as the extended third motion shaft must be lined up with the splines of the uni-directional clutch.

The unit is split at the adapter plate which is attached to the front casing by six $\frac{5}{16}$ in. (7.938 mm.) studs, two of which are extra long. The four nuts on the shorter studs should be removed before those on the longer ones are touched. The latter should be unscrewed

together releasing the compression of the clutch springs The unit can then be drawn off the mainshaft, leaving the adapter plate attached to the gearbox.

Remove the clutch springs from their pins. The two bridge pieces against which the operating pistons bear can now be removed. Each is secured by two $\frac{1}{4}$ in. nuts locked by tab washers. Withdraw the two operating pistons.

As the adapter plate is now separated from the unit the pump valve can be dismantled without removing the side cover (solenoid bracket) from the casing and there is no need to disturb the latter unless it is necessary to remove the accumulator piston and spring.

Remove the six $\frac{5}{16}$ in. (7.938 mm.) nuts securing the two halves of the casing and separate them, removing the brake ring which is spigoted into the two pieces.

Lift out the planet carrier assembly. Remove the clutch sliding member complete with the thrust ring and bearing, the sun wheel and thrust washers. Take out the inner member of the uni-directional clutch, the rollers, cage, etc.

If it is necessary to remove the planet gears from the carrier the three split pins securing the planet bearing shafts must be extracted before the latter can be knocked out.

To remove the annulus, first take off the coupling flange at the rear of the unit, remove the speedometer gear, centrifugal switch, etc., and drive out the annulus from the back. The front bearing will come away on the shaft leaving the rear bearing in the casing.

INSPECTION

Each part should be thoroughly inspected after the unit is dismantled and cleaned to ensure which parts should be replaced. It is important to appreciate the difference between parts which are worn sufficiently to affect the operation of the unit and those which are merely "worn in."

1. Inspect the front casing for cracks, damage, etc. Examine the bores of the operating cylinders and accumulator for scores and wear. Check for leaks from plugged ends of the oil passages. Ensure that the welch washer beneath the accumulator bore is tight and not leaking. Inspect the support bushes in the centre bore for wear and damage.

2. Examine the clutch sliding member assembly. Ensure that the clutch linings are not burned or worn. Inspect the pins for clutch springs and bridge pieces and see that they are tight in the thrust ring and not distorted. Ensure that the ball bearing is in good condition and rotates freely. See that the sliding member slides easily on the splines of the sun wheel.

3. Check the clutch springs for distortion or collapse.

4. Inspect the teeth of the gear train for damage. If the sun wheel or planet bushes are worn the gears will have to be replaced since it is not possible to fit new bushes in service because they have to be bored true to the pitch line of the teeth.

5. Examine steel and bronze thrust washers.

6. See that the rollers of the uni-directional clutch are not chipped and that the inner and outer members of the clutch are free from damage. Make sure that the member is tight in the annulus. Ensure that the spring is free from distortion.

7. Inspect the ball bearings on the output shaft and see that there is no roughness when they are rotated slowly.

8. Ensure that there are no nicks or burrs on the mainshaft splines and that the oil holes are open and clean.

9. Inspect the oil pump for wear on the pump plunger and roller pin. Ensure that the plunger spring is not distorted. Free length 2 in. (5.08 cm.). Inspect the valve seat and ball and make sure that they are free from nicks and scratches.

10. Check the operating valve for distortion and damage and see that it slides easily in its bore in the front casing.

REASSEMBLING THE UNIT

The unit can be reassembled after all the parts have been thoroughly cleaned and checked to ensure that none are damaged or worn.

Assemble the annulus into the rear casing, not forgetting the spacing washer which fits between a shoulder on the shaft and the rear ball bearing. This washer is available in different thicknesses for selective assembly and should allow no end float of the annulus (output shaft) and no pre-loading of the bearings.

Selective washers are furnished in the following sizes:—
.146 in. ± .0005 in. (3.7084 mm. ± .0127); .151 in. ± .0005 in. (4.335 mm. ± .0127); .156 in. ± .0005 in. (3.962 mm. ± .0127); .161 in. ± .0005 in. (4.089 mm. ± .0127).

Replace the thrust washer and uni-directional clutch inner member with its rollers and cage. The fixture (Fig. 10) is for retaining the rollers in position when assembling the clutch. Ensure that the spring is fitted correctly so that the cage urges the rollers up the ramps on the inner member.

Fit the pump cam on to the gearbox mainshaft, offer up the front housing to the cover plate and secure temporarily with two nuts. In order to determine the amount of end float of the sun wheel, which should be .008 in.–.014 in. (.203–.3556 mm.) an extra thrust washer of known thickness should be assembled with the two normally used in front of the sun wheel.

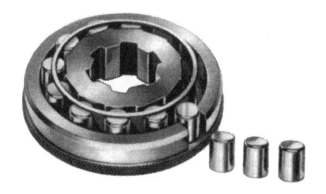

Fig. 10. Using tool 18G 178 for assembling the roller clutch.

Fit the planet carrier, with its planet gears over the sun wheel, and with the assembly in this position offer it up to the annulus. Turn the planet carrier until the locating peg on the inner member of the uni-directional clutch enters the corresponding hole in the planet carrier. This lines up the splines in the two members.

Assemble the brake ring to the front casing then offer up the front and rear assemblies, leaving out the clutch sliding member with its springs, etc. The gap between the flanges of the brake ring and rear casing should be measured. This gap will be less than the thickness of the extra thrust washer by the amount of end float of the sun wheel. If this is between the limits

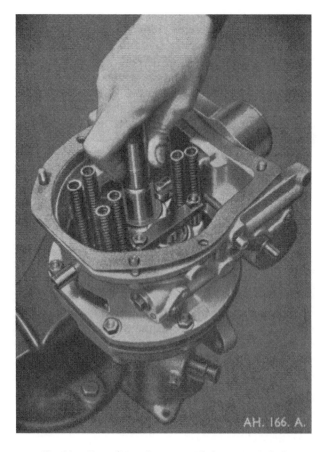

Fig. 11. *Centralising the gears with dummy mainshaft.*

specified the unit may be stripped down again and re-assembled without the extra thrust washer. The clutch sliding member bridge pieces, etc., must now be replaced. The compression of the springs is taken up on the two long studs between the front casing and adaptor plate.

If the indicated end float is more, or less, than that required it must be adjusted by replacing the steel thrust washer at the front of the sun wheel by one of less or greater thickness, as required. Washers of varying thickness are stocked for this purpose.

Seven sizes are available, as follows :—

.113 in.–.114 in. (2.87 mm.-2.985 mm.); .107 in.–

.108 in. (2.718 mm.-2.743 mm.); .101 in.–.102 in. (2.565-2.59 mm.) ; .095 in.–.096 in. (2.413 mm.-2.438 mm.); .089 in.–.090 in. (2.26 mm.-2.286 mm.); .083 in.–.084 in. (2.108-2.134 mm.); .077 in.–.078 in. (1.956 mm.-1.981 mm.).

Care must be taken to ensure that the thrust washers at the front and rear of the sun wheel are replaced in their correct positions. At the front of the sun wheel the steel washer fits next to the head of the support bush in the housing and the bronze washer between the steel one and the sun wheel. At the rear, the steel washer is sandwiched between the two bronze washers.

Grip the mounting flange of the overdrive unit in a vice, so that the unit is upright, and insert a dummy shaft 18G 185 or a spare mainshaft if the dummy shaft is not available, so that the sun wheel and thrust washers, planet carrier and roller clutch line up with each other; a long thin screwdriver should be used to line by eye the splines in the planet carrier and the roller clutch before inserting the dummy shaft. Gently turn the coupling flange to and fro while holding the dummy shaft, to assist in feeling the shaft into the splines of the planet carrier and roller clutch. Make sure that the dummy shaft has gone right in by holding the coupling flange in one hand and turning the shaft to and fro to feel the free-wheel action of the roller clutch.

Make quite sure that the clutch springs are in their correct positions—the $4\frac{1}{4}''$ (10.8 cm.) long springs are the inner ones, and the $4\frac{1}{2}''$ (11.5 cm.) ones are the outer. This is most important because if any of the springs are in the wrong position they will become "coil bound" when the adapter plate is in place and restrict the movement of the sliding clutch so that overdrive will not engage.

Place the oil pump operating cam in position on top of the centre bushing with the lowest part of the cam in contact with the oil pump plunger and also place the paper joint washer in position.

The gearbox, with top gear engaged, should now be lifted by hand on to the overdrive unit, carefully threading the mainshaft through the oil pump cam and into the centre bushing in the body of the overdrive unit. Gently turn the first motion shaft to and fro to assist in "feeling" the mainshaft into the splines of the planet carrier. When the mainshaft is sufficiently entered for the gearbox to come to rest against the clutch springs with the two long studs just protruding through the holes in the overdrive body, put the spring washers and nuts on to the end of the studs. Before commencing to tighten the nuts, use a long thin screwdriver to guide the ends of the clutch springs on to the short locating pegs which are cast into the face of the adapter plate—this is very important because if the springs are not properly located they may become "coil bound" and prevent overdrive engaging. Now commence simultaneously to tighten the nuts on the two long studs, compressing the clutch springs and drawing the gearbox and overdrive together evenly. As the gearbox and overdrive come together watch carefully to see the splines on the mainshaft enter the oil pump operating cam and that the cam remains properly engaged with the oil pump plunger. If the two units do not pull together easily with only the resistance of the clutch springs being felt as the two nuts are tightened, stop tightening immediately. Gently rotate the gearbox first motion shaft in a clockwise direction whilst holding the overdrive coupling flange stationary until the mainshaft is felt to enter the roller clutch. The tightening of the nuts on the two long studs can then be completed, and the nuts fitted and tightened on to the four short studs. NOTE: The gearbox mainshaft should enter the overdrive easily, provided that the lining-up procedure previously described is carried out and the unit is not disturbed. If any difficulty is experienced it is probable that one of the components has become misaligned, and the gearbox should be removed and the overdrive re-aligned with the dummy shaft.

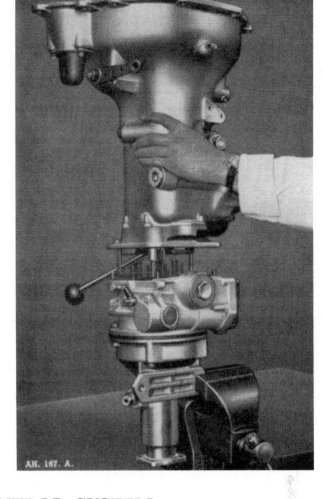

Fig. 12. *With the overdrive held firmly in a vice and the dummy main shaft removed after lining up, lower the gearbox into place.*

TWO-RELAY CONTROL SYSTEM

Before a change into overdrive can be made, three conditions must first be satisfied. These are:—

(i) The manual switch to be closed by driver.

(ii) The car to be in top or second gear.

(iii) The road speed to have reached a minimum of approximately 40 m.p.h. (64 k.p.h.)

When these conditions exist, the operating coil of relay R1 will be energised and its contacts will close.

Relay R1 closing will connect to the supply, the solenoid and the operating coil of relay R2.

The solenoid, being energised, actuates a soft iron plunger which in turn lifts the operating valve of the overdrive unit, thereby effecting a change into overdrive.

RELAY R2 AND THROTTLE SWITCH

It is important that a return from overdrive to top or second gear is not made with a slow running engine and a fast road speed, and the purpose of relay R2 is to make a return under these conditions impossible.

The operating coil of relay R2 is connected in series with a throttle switch the contacts of which are closed at any throttle opening below approximately one fifth. The contacts of relay R2 are connected in parallel with the contacts of the centrifugal and manual switches.

Relay R2 will therefore maintain the coil circuit of relay R1 in the event of the centrifugal switch contacts opening or the manual switch being turned off at throttle openings less than one fifth. At above approximately one fifth throttle, overdrive can be selected or switched off at will by means of the manual switch, providing the road speed is high enough to cause the centrifugal switch contacts to close.

Should a period of fast driving in overdrive be followed by a period of coasting with the throttle closed, the car will decelerate and the centrifugal switch contacts will open, but the transmission will continue to be through overdrive because the throttle switch contacts will have closed.

If, now, the throttle is slowly depressed, power will steadily and increasingly be applied from the engine to the road wheels. The centrifugal switch contacts

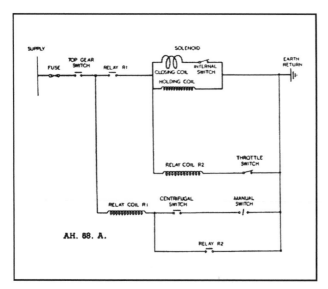

Fig. 13. *Wiring diagram for the overdrive electrical circuit.*

Should, however, the throttle be depressed rapidly beyond about a fifth open, the throttle switch contacts will have separated before the centrifugal switch has closed. The overdrive will therefore disengage and remain disengaged until the car has attained a speed sufficient to cause the centrifugal switch contacts to reclose.

OPERATING SOLENOID

The solenoid is made up of a soft iron plunger, a holding coil of high resistance, a closing coil of low resistance and a pair of normally closed contacts. These contacts are connected in series with the closing coil.

When the contacts of relay R1 close, both coils in the solenoid become energised and actuate the soft iron plunger. Movement of the plunger opens the solenoid internal switch and cuts out the low resistance closing coil, the magnetism due to the high resistance coil alone being sufficient to keep the plunger in the overdrive position.

should therefore close before the throttle switch contacts have opened, and overdrive will be maintained.

TRACING DEFECTS IN THE TWO-RELAY SYSTEM

SYMPTOM No. 1

Overdrive Fails to Operate

(a) Check fuse to which overdrive circuit is connected. If fuse has blown, disconnect the overdrive supply cable at fuse terminal and fit new fuse. Operate each of the components protected by it. If the fuse again blows, the fault is in one of the external circuits and in the overdrive circuit.

If the fuse does not blow, examine carefully all wiring of the overdrive circuit for signs of short-circuits or loose connections, before reconnecting the overdrive supply cable.

(b) If the fuse has not blown and the wiring is in order then, with the ignition switched on and the engine stopped, check as follows :

(i) Engage top gear and short out terminals C1–C2 of relay R1.

The overdrive solenoid should operate. If no operation takes place, the top gear switch may be defective. This can be checked by linking terminal C1 of relay R1 to the fuse, using a suitable length of insulated cable. Operation of the solenoid indicates a defective top and second gear switch. Remove test link and fit new gear switch.

(ii) Engage top gear and short out terminals C1–C2 of relay R2. The overdrive solenoid should operate. If no operation takes place, either the coil or the contacts of relay R1 may be defective.

This may be checked by shorting out terminals C1–C2 of relay R1. Operation of overdrive solenoid indicates a defective relay R1. Disengage top gear and fit new relay.

(iii) Engage top gear, switch on manual switch and short out centrifugal switch terminals.

The overdrive solenoid should now operate, thus indicating a faulty centrifugal switch.

Warning. Any movement of the car in a forward direction with the centrifugal switch loose or removed will result in the speedometer driven gear being displaced. Ensure that this gear is correctly fitted when replacing the centrifugal switch.

(iv) Failure of solenoid to operate in above test may indicate a faulty manual switch. To check this, short out terminals C1–C2 of relay R2. Operation of the overdrive solenoid indicates a faulty manual switch.

(v) If the solenoid fails to operate in test (iv) the solenoid is faulty and should be replaced by a new unit.

SYMPTOM No. 2

Action of Overdrive Solenoid Weak. Solenoid is defective and should be replaced.

SYMPTOM No. 3

Overdrive Operates, but Solenoid Unit Overheats. Solenoid is defective and should be replaced.
Note:—The normal current consumption of the system should be approximately 1.0–1.5 amperes. The above fault will be accompanied by a current of approximately 18–20 amperes.

SYMPTOM No. 4

Overdrive drops out with throttle closed, accompanied by noticeable braking effect.

(a) Check setting of throttle switch, linkage and, if necessary, reset.

(b) Check throttle switch contacts as follows :—
Switch on ignition, engage top gear, close throttle, short out terminals C1–C2 of relay R1, the solenoid should now operate. If it does not, link terminal of relay R2 to earth. Operation of overdrive will indicate a faulty throttle switch.

(c) Check for faulty operation of relay R2 contacts as follows :—

(i) Connect a 0–20 voltmeter to terminals C1–C2 of relay R2, and engage top gear. A reading of approximately 12 volts should be obtained.

(ii) Short out terminals C1–C2 of relay R1. If previous checks have been satisfactory, the overdrive will operate and the voltmeter reading should drop immediately to zero. If any reading at all is obtained, replace the relay with a good unit.

SERVICE JOURNAL REFERENCE

NUMBER	DATE	SUBJECT	CHANGES

PROPELLER SHAFT

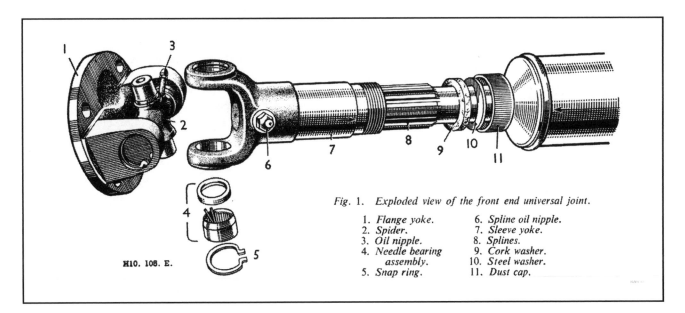

Fig. 1. Exploded view of the front end universal joint.

1. Flange yoke.	6. Spline oil nipple.
2. Spider.	7. Sleeve yoke.
3. Oil nipple.	8. Splines.
4. Needle bearing assembly.	9. Cork washer.
	10. Steel washer.
5. Snap ring.	11. Dust cap.

H10. 108. E.

Description

The Propeller Shaft and Universal Joints (Fig. 1) are of Hardy Spicer manufacture.

The fore and aft movement of the rear axle and other components is allowed for by a sliding spline between the propeller shaft and gearbox unit. Each universal joint consists of a centre spider, four needle roller bearings and two yokes. Reference to the Lubrication Chart on page S/1 shows the location of the joints.

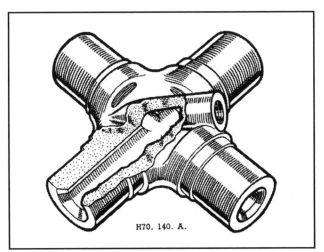

H70. 140. A.

Fig. 2. Showing oil channels in a joint spider.

Lubrication

An oil nipple is fitted to each centre spider for the lubrication of the bearings. Grease must not be used, oil being the correct lubricant. Reference to Fig. 2

shows that the central oil chamber is connected to the four oil reservoirs and to the needle roller bearing assemblies.

The needle roller bearings are filled with oil on assembly. An oil nipple is provided on the sleeve yoke of the sliding spline joint for lubrication of the splines.

If a large amount of oil exudes from the oil seals the joint should be dismantled and new oil seals fitted.

After dismantling, and before reassembly, the inside splines of the sleeve yoke should be smeared liberally with oil.

Tests for Wear

Wear on the thrust faces is located by testing the lift in the joint, either by hand, or by using a length of wood suitably supported.

Any circumferential movement of the shaft relative to the flange yokes, indicates wear in the needle roller bearings, or the sliding spline.

Removal of Complete Assembly

Before removal of the propeller shaft can be effected, the short length of tunnel immediately to the rear of the gearbox must be removed.

The removal procedure for the propeller shaft is as follows :—

Support the shaft near the sliding joint, then withdraw the bolts from the gearbox companion flange.

Unscrew, by hand, the dust cap at the rear of the sliding joint. Slide the splined sleeve yoke about half an inch rearwards, thus disengaging the pilot flanges.

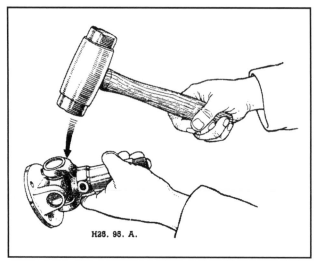

Fig. 3. Tapping the joint to extract bearing.

Next remove the four nuts and bolts securing the rear flange yoke from the axle companion flange and lower the propeller shaft to the ground.

The propeller shaft and the two universals can now be taken to the bench for further dismantling.

Dismantling

The following directions apply to both universal joints of the propeller shaft except for the fact that the front joint can be separated from the shaft, whereas the rear joint has one yoke permanently fixed to the tube.

Clean away the enamel from all the snap rings and bearing faces, to ensure easy extraction of the bearings.

Remove the snap rings by pressing together the ends of the rings and extract with a screwdriver. If the ring

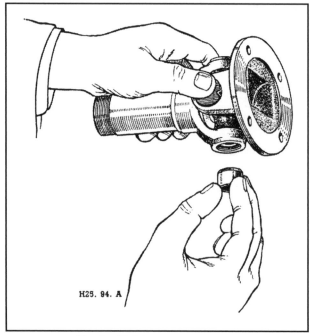

Fig. 4. Withdrawing a bearing cup.

does not come out easily, tap the bearing face lightly to relieve the pressure against the ring.

Hold the splined end of the shaft in one hand and tap the radius of the yoke with a lead or copper hammer (see Fig. 3), when it will be found that the bearing will begin to emerge. If difficulty is experienced, use a small bar to tap the bearing from the inside, taking care not to damage the race itself. Turn the yoke over and extract the bearing with the fingers (see Fig. 4), being careful not to lose any of the needles.

Repeat this operation for the other bearing, and the splined yoke can be removed from the spider (see Fig. 5).

Using a support and directions as above remove the spider from the other yoke.

Examination and Checking for Wear

After long usage the parts most likely to show signs of wear are the bearing races and the spider journals of the universal joints. Should looseness or stress marks

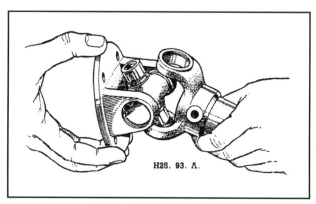

Fig. 5. Separating the joint.

be observed, the assembly should be renewed complete, as no oversize journals or bearings are provided.

It is essential that bearing races are a light drive fit in the yoke trunnions. Should any ovality be apparent in the trunnion bearing holes, new yokes must be fitted.

With reference to wear of the cross holes in a fixed yoke, which is part of the tubular shaft assembly, only in cases of emergency should this be replaced. It should normally be renewed with a complete tubular shaft assembly. The other parts likely to show signs of wear are the splined sleeve yoke, or splined stub shaft. A total of .004 in. circumferential movement, measured on the outside diameter of the spline, should not be exceeded. Should the splined stub shaft require renewing, this must be dealt with in the same way as the fixed yoke, i.e. a replacement tubular shaft assembly fitted.

Reassembly

See that all drilled holes in the journals of the universal joints are cleaned out and filled with oil (Fig. 2). Assemble the needle rollers in the bearing races and fill with oil. Should difficulty be experienced in assembly, smear the walls of the races with vaseline to retain the needle rollers in place.

Insert the spider in the flange yoke. Using a soft-nosed drift about $\frac{1}{32}$ in. smaller in diameter than the hole in the yoke, tap the bearing in position. It is essential that bearing races are a light drive in the yoke trunnion. Repeat this operation for the other three bearings. The spider journal shoulders should be coated with shellac prior to fitting the retainers to ensure a good seal.

If the joint appears to bind, tap lightly with a wooden mallet which will relieve any pressure of the bearings on the end of the journals. When replacing the sliding joint on the shaft, be sure that the trunnions in the sliding and fixed yoke are in line. This can be checked by observing that arrows marked on the splined sleeve yoke and the splined stub shaft are in line. It is advisable to renew cork washers and washer retainers on spider journals, using a tubular drift.

Replacing the Shaft Assembly

Wipe the companion flange and flange yoke faces clean, to ensure that the pilot flange registers properly and the joint faces bed evenly all round. Insert the bolts, and see that the nuts are tightened evenly all round and are securely locked.

The dust cap must be screwed up by hand as far as possible. The sliding joint is always placed towards the front of the car.

SERVICE JOURNAL REFERENCE

NUMBER	DATE	SUBJECT	CHANGES

FRONT HUBS AND INDEPENDENT FRONT SUSPENSION

GENERAL DATA

Hub Bearings :

    Inner : R & M LJ 1¼, Double Purpose Ball Journal (Light) Size 1¼ × 2¾ × 11/16 in. (3.175 × 6.985 × 1.746 cm.)

    Outer : R & M MJ ¾, Double Purpose Ball Journal (Medium), Size ¾ × 2 × 11/16 in. (1.905 × 5.081 × 1.746 cm.)

Castor Angle	1¾°
Camber Angle	1°
Swivel Pin Inclination	6½°

Swivel Pin Thrust Bearing :

                              Oilite washer between two stainless steel washers.

Swivel Pin Diameter :

Top	.686¼ to .686¾ in. (17.43 to 17.44 mm.)
Bottom	.811¼ to .811¾ in. (20.60 to 20.61 mm.)

Swivel Pin Bush Tolerance :

Top	.000¾ to .001¾ in. (.019 to .04 mm.)
Bottom	.000¾ to .001¾ in. (.019 to .04 mm.)

Independent Front Spring :

Free Length	11.14 in. (28.04 cm.)
Fitted Length	7 in. (17.78 cm.)
Number of Effective Coils	7
Diameter of Wire	0.531 in. (13.48 mm.)
Inside Diameter of Coil	3.589 in. (9.11 cm.)

Shock Absorbers :

Make	Armstrong Hydraulic
Type	Double Acting IS9/10R IS9/10RXP

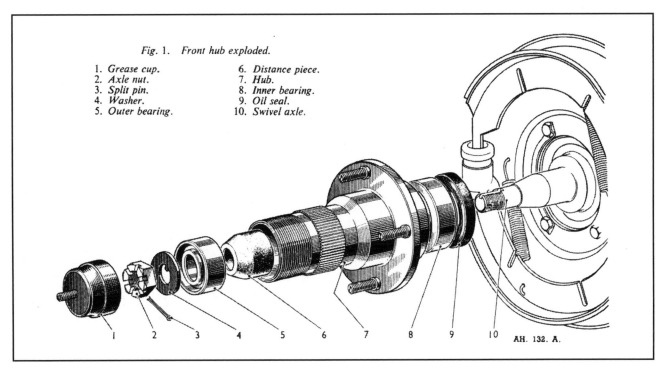

Fig. 1.   Front hub exploded.

1. Grease cup.
2. Axle nut.
3. Split pin.
4. Washer.
5. Outer bearing.
6. Distance piece.
7. Hub.
8. Inner bearing.
9. Oil seal.
10. Swivel axle.

AH. 132. A.

FRONT HUBS

Fig. 2. Using the special extractor to remove the grease retaining cup.

To Check for Wear

The inner and outer ball bearings of the front hub are non-adjustable, the amount of thrust being determined by a distance piece. To check for wear of these bearings the car should be jacked up until the wheel of the front hub to be checked, is clear of the ground. Movement between the wheel and the backplate denotes wear of the hub bearings. Should a very positive movement be apparent, the front hub bearings will need renewing.

Dismantling the Front Hubs

To dismantle either front hub, first jack up the car until the wheel is clear of the ground and then place blocks under the independent spring plate. Lower the car on to the blocks.

Remove the "knock-on" hub cap (direction of rotation marked on cap) and pull the wheel off the splines. Release the nuts and washers holding the brake drum, then gently tap the brake drum clear of the front hub assembly. If the drum appears to bind on the brake shoes, the shoe adjusters should be slackened.

Use the extractor provided in the tool kit to extract the grease retaining cup from within the hub, see fig. 2. Straighten the end of the split-pin and then prize it out through the hole provided in the hub. Using a box spanner and tommy bar remove the hub securing nut and flat washer from the swivel axle.

The front hub can now be withdrawn by using an extractor. It is preferable to use an extractor which screws into position on the hub cap thread, but an ex-

tractor which locates over the hub studs may also be used. The hub is withdrawn complete with the inner and outer bearings, the distance piece and oil seal. Should the inner bearing remain on the swivel axle it can be removed by inserting a thin rod in turn into the two small holes in the back of the swivel axle, either side of the swivel pin, and tapping the race lightly. If an extractor is used to remove the race there is a danger that the outer ring of the race will be pulled clear of the balls and the bearing will fall apart.

With the hub removed, the outer bearing and distance piece can be dismantled by inserting a drift through the inner bearing and gently tapping the outer bearing clear of the hub. The inner bearing and oil seal can then be removed by inserting the drift from the opposite side of the hub.

The removal of the brake backplate is described in Section M.

Assembling the Front Hubs

When assembling the hub the inner ball bearing race should first be inserted into the hub with the side of the race marked "thrust" facing the distance piece.

Pack the hub with *recommended* grease and then insert the distance piece so that the smaller end faces the outer bearing. Never use a grease thicker than that recommended.

Replace the outer bearing so that the "thrust" side faces the distance piece. Use a soft metal drift to replace both bearings tapping them gently on diametrically opposite sides to ensure that they move evenly into their

Fig. 3. Showing the grease cup removed.

1. Extractor.   2. Grease cup   3. Hub.

respective housings in the hub. Replace the hub oil seal over the inner bearing so that the hollow side of the seal faces the bearing. Renew the seal if it is damaged in any way.

The hub can now be replaced on the swivel axle. This is best done by using a hollow drift which will bear evenly on both the inner and outer races of the outer hub bearing. Gently tap the hub into position until the inner race bears against the shoulder on the swivel axle.

Place the flat washer on the swivel axle and screw the nut down finger tight with the aid of a box spanner. Turn the hub and note the resistance, which at this stage is due to the oil seal. Then continue tightening the nut until a slightly increased resistance to the turning of the hub is noticed. The bearings are now pre-loaded and the split pin should be inserted to lock the nut. To do this, rotate the hub until the hole in the bearing is in line with the hole in the swivel axle.

Pack the retaining cup with grease and, using a drift, tap it gently but firmly up against the outer bearing.

Replace the brake drum and secure with the four spring washers and self-locking nuts. Before refitting the wheel, the spline and cone faces of both the hub and wheel, also the threads of the hub cap should be smeared with grease.

Fig. 4. Extracting the front hub.
1. Extractor.   2. Adapter.   3. Hub.

Refit the wheel, screw on the hub cap and tighten with the aid of a hide hammer. Before lowering the car to the ground and finally tightening the hub cap, make sure that the brake shoes are correctly adjusted.

INDEPENDENT FRONT SUSPENSION

Description

The independent front suspension is known as the "Wishbone" type, since the top and bottom linkages roughly conform to the shape of a wishbone. Between these two wishbones is the coil spring, held under compression between the top spring plate which is welded to the chassis side member, and the lower spring plate which is secured to the lower wishbone by four bolts.

The top wishbone is secured at the chassis end to a double-acting hydraulic shock absorber which is secured to the top spring plate bracket by four setpins. The two arms of the top wishbones thus form the operating levers of the shock absorber. At the swivel end, the top wishbone is secured to the swivel pin trunnion by means of a fulcrum pin and tapered rubber bushes. The bottom wishbone is secured by fulcrum pins and tapered rubber bushes to two brackets on the chassis frame and by two screwed bushes and a screwed fulcrum pin to the lower end of the swivel pin.

Anti-Roll Torsion Bar

The anti-roll bar is held to each chassis dumb-iron by two setpins and connected to either suspension unit by a short link rod, secured by a self-locking nut at each end.

Checking for Wear

The following parts of the independent front suspension are liable to wear. Rectification may mean the fitting of new parts or assemblies.

Swivel Pin and Bushes: Wear of the swivel pin, or bushes, or both, may be checked by jacking up the front of the car and endeavouring to rock the wheel by grasping opposite points of the tyre in a vertical position. If any sideways movement can be detected between the swivel axle assembly, the swivel pin or the swivel pin bushes are worn and must be stripped for examination.

Shock Absorber Bearings: Up and down, or sideways movement of the shock absorber cross shaft, relative to the shock absorber casting, denotes wear of the shock absorber shaft bearings which can only be remedied by refitting a new shock absorber. These bearings are best checked when the suspension is dismantled and when with some freedom of movement, it is possible to move the top wishbone arms, which are attached at their inner ends to the shock absorber cross shaft.

Wishbone Arm Rubber Bushes: The rubber bearing bushes used for the upper wishbone arm outer bearings and for the lower wishbone arm inner bearings may in time deteriorate and need renewing. Excessive sideways movement in either of these bearings would denote softening of the rubber bushes.

Wishbone Arm Screwed Bush Bearing: The screwed bushes or the screwed trunnion fulcrum pin of the lower wishbone arm outer bearing assembly may develop excess free play due to wear of either of these parts. This assembly can best be checked when the suspension has been dismantled.

Removing the Coil Spring

Slacken the "knock-on" hub cap of the wheel concerned, then place a jack under the chassis front cross member and raise the car until the front wheels are clear of the ground. Unscrew the cap and remove the wheel.

In the absence of Service Tool 18G 37, two $\frac{3}{8}$ in. B.S.F. slave bolts will be required to release the compression from the coil spring. These bolts should be of high-tensile steel, 4 in. long and threaded their entire length.

There are four nuts and bolts securing the bottom spring plate to the suspension lower links, the nuts being of the self-locking type. Unscrew the nuts from two diagonally opposite bolts. Remove these bolts and insert the two slave bolts in the vacated holes. Screw their nuts down securely then remove the two remaining short bolts. Unscrew the nuts from the slave bolts, each, a little at a time. When the spring is fully extended, release the bolts and remove the spring plate and coil spring.

Fig. 5. Illustrating the coil spring compressor (Service Tool 18G 37) in position.

Checking the Spring

The spring should be checked for correct free length as given on page J/1. The spring should be renewed if there is any variation in its correct length.

Removing the Suspension

Jack up the car, remove the wheel and coil spring as already explained. Disconnect the steering side-tube from the steering arm by withdrawing the split pin and unscrewing the nut. Also disconnect the flexible brake pipe from the brake backplate, tying it to some higher point to prevent unnecessary loss of fluid.

With the suspension unit supported, remove the fulcrum pins securing the lower wishbone arms to their brackets on the frame, taking care to retrieve the two rubber bushes and special washer from each bearing. Unscrew the four setpins securing the shock absorber to the top spring bracket. The suspension unit is now free to be lifted clear.

Dismantling the Suspension

The top wishbone arms are connected at their narrowest point by a clamping bolt. Unscrew the nut and release this bolt. Next remove the split pin and nut from the upper trunnion fulcrum pin on the outer end of the top wishbone arms.

The forward arm (left-hand suspension unit) of the top wishbone is secured to the shock absorber spindle by a clamping bolt. Slacken the clamping bolt and partially withdraw the arm. The trunnion fulcrum pin can now be withdrawn and the shock absorber removed complete with the top wishbone arms.

Withdraw the rubber bearing from each end of the upper trunnion. These bearings fit into a groove in the swivel pin and *must* be taken out before the swivel pin can be removed. Take out the split pin and unscrew the nut from the top of the swivel pin. Remove the upper trunnion and the three thrust washers and lift off the swivel axle and hub assembly. Detach the cork washer from the lower end of the swivel pin.

The outer bearing of the lower wishbone arms can now be dismantled. Slacken the nut on each of the half-moon cotters located in the ends of the lower wishbone arms, screw out the two threaded bushes and detach the arms.

Unscrew the nut from the cotter, located in the centre of the lower trunnion, and tap out the cycle-type cotter. Withdraw the fulcrum pin and remove the cork washer from each end of the trunnion.

The suspension unit is now dismantled and worn or damaged parts can be renewed.

Examination for Wear
Swivel Pin: If wear of the swivel pin and bushes is suspected as described earlier, carefully examine the swivel pin for wear by checking for ovality with a

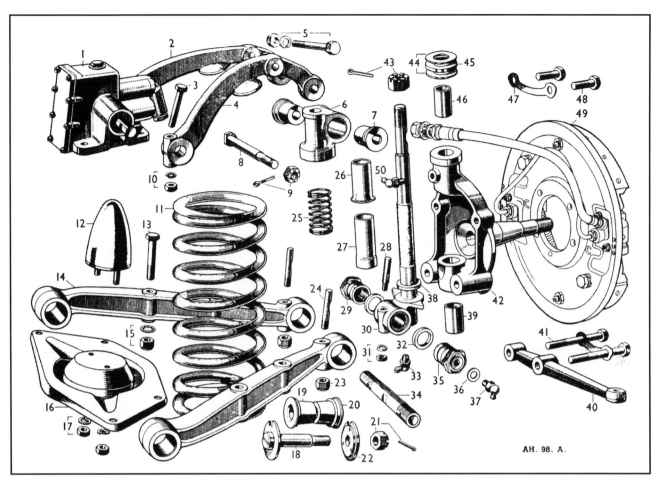

Fig. 6.   *Components of the independent front suspension.*

1 *Shock absorber.*	18. *Fulcrum pin for inner lower bearing.*	35. *Front screw bush.*
2. *Rear top wishbone arm.*	19. *An inner lower rubber bearing.*	36. *Flat washer.*
3. *Clamping bolt for front wishbone arm.*	20. *An outer lower rubber bearing.*	37. *Oil nipple.*
4. *Front top wishbone arm.*	21. *Fulcrum pin nut and split pin.*	38. *Cork ring.*
5. *Joining bolt for top wishbone arms.*	22. *Fulcrum pin special washer.*	39. *Swivel axle lower bush.*
6. *Upper trunnion link.*	23. *Nut for bush cotter.*	40. *Steering arm.*
7. *Trunnion rubber bearing.*	24. *Bush cotter.*	41. *Steering arm setpin.*
8. *Upper trunnion fulcrum pin.*	25. *Swivel pin dust cover spring.*	42. *Swivel axle.*
9. *Fulcrum locking nut and split pin.*	26. *Upper dust cover.*	43. *Swivel pin nut and split pin.*
10. *Nut and washer for clamping bolt.*	27. *Lower dust cover.*	44. *Staybrite washers.*
11. *Coil spring.*	28. *Cotter for fulcrum pin.*	45. *Oilite washer.*
12. *Rebound rubber bumper.*	29. *Rear screwed bush.*	46. *Swivel axle upper bush.*
13. *Spring plate bolt.*	30. *Swivel pin and lower trunnion.*	47. *Back plate setpin lockwasher.*
14. *Rear lower wishbone arm.*	31. *Nut and washer.*	48. *Back plate setpin.*
15. *Simmonds nut and lockwasher.*	32. *Cork ring.*	49. *Back plate assembly.*
16. *Spring plate.*	33. *Trunnion oil nipple.*	50. *Swivel pin oil nipple.*
17. *Rebound bumper nut and washer.*	34. *Screwed fulcrum pin.*	

micrometer gauge. Should the pin not show any appreciable signs of wear renewal of the swivel bushes may effect a satisfactory cure. These bushes can easily be driven out and replaced with a suitable drift. When refitting the top bush the oiling hole must locate with the oil hole in the swivel housing and the top of the bush must be flush with the swivel housing. The second bush must be flush with the recessed housing and protrude about ⅛ of an inch above the lower housing upper face. Then ream the bushes from the bottom as necessary with Service Tools 18G 64 and 18G 65.

The two-piece dust cover for the swivel pin is easily removed and replaced by telescoping the spring loaded dust cover tubes.

Wishbone Arm Screwed Bush Bearing: If it is found that the screwed bushes can be moved backwards or forwards on the fulcrum pin thread they should be renewed. Should new screwed bushes on the old fulcrum pin still permit end play, then renew the fulcrum pin.

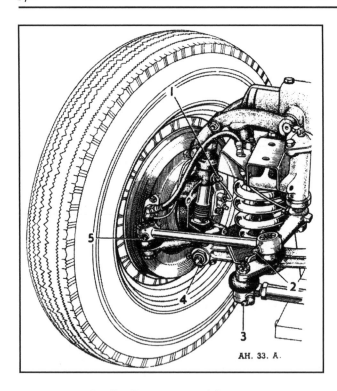

AH. 33. A.

Fig. 7. Front suspension lubrication.

1. *Swivel pin.*   2. *and 5. Side rod connections.*
3. *Cross tube connection.*   4. *Lower link bushes.*

Shock Absorbers : The cross shaft bearings of the double acting hydraulic shock absorber may have worn sufficiently to permit up and down or sideways movement of the cross shaft. If such wear is apparent the shock absorber must be renewed complete.

The shock absorber should also be carefully examined for any leaks and tested for effective damping. Secure the shock absorber mounting plate in a vice and move the wishbone arms up and down through a complete stroke. A moderate resistance throughout the full stroke should be felt.

If resistance is erratic it may mean that the fluid level is too low and that there are air locks in the shock absorber. To rectify this, the shock absorber filler plug should be removed and the fluid level maintained at the correct level while the arms are moved steadily up and down through full strokes. If this treatment does not effect a cure the shock absorber must be renewed as a complete unit.

The shock absorbers cannot be overfilled because the deep filler opening prevents all the air from escaping. Top up until the fluid begins to rise in the filler.

Assembling the Suspension

First fit the screwed fulcrum pin into the lower trunnion at the bottom end of the swivel pin, ensuring that it is centralized and secured by means of the cycle type cotter. Fit a cork ring into the recess provided at each end of the lower trunnion and introduce the lower wishbone arms into position. Ensure that the half-moon cotters are correctly positioned to receive the steel bushes which should now be greased and screwed partially home.

To ensure that the alignment of the lower wishbone arm is correct, it is necessary, in the absence of Service Tool 18G 56, to bolt the lower spring plate securely in position. Screw the threaded bushes home evenly, then slacken them back one flat. Finally secure the bushes by tightening the nuts on each of the half-moon cotters. Do not overtighten the cotter nuts as this may cause distortion of the bushes. If the assembly has been correctly carried out it will be possible to insert a .002 in. feeler gauge between the inner shoulder of the bush and the outer face of the wishbone arm on each side. The lower trunnion assembly should now operate freely in the screwed bushes.

Place the cork washer on the swivel pin with its chamfered face downwards and smear the swivel pin with a little clean engine oil. Then position the swivel axle and hub assembly on the swivel pin. The thrust washers, comprising an "Oilite" washer interposed between two "Staybrite" washers should next be fitted. The "Staybrite" washers are supplied in varying thicknesses to permit adjustment, as it is necessary to provide easy operation of the swivel axle with the minimum amount of lift. The maximum permissible lift is .002 in. Fit the upper trunnion and swivel nut, and check the clearance, correcting it if necessary by means of the "Staybrite" washers. Then slacken the swivel pin nut to permit further assembly.

Moisten the upper trunnion rubber bearings with water and place them in position. The trunnion with its bearings should next be placed in position between the two upper wishbone arms, after which the fulcrum pin

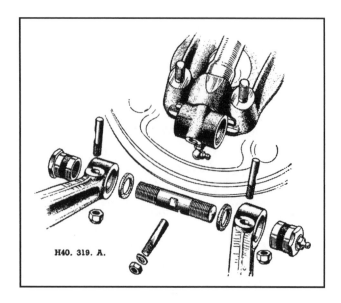

H40. 319. A.

Fig. 8. This exploded view shows the screwed bush housing assembly at the lower end of the wishbone arms.

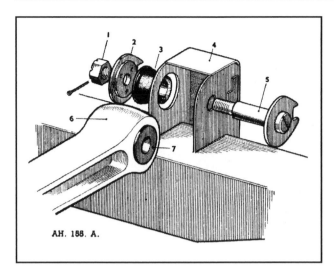

AH. 188. A.

Fig. 9.   Lower wishbone inner bearing assembly.

1. Castellated nut.	*4. Mounting bracket.*	*6. Wishbone*
2. Special washer.	*5. Fulcrum pin.*	*arm.*
3. Bush.		*7. Bush.*

should be fitted and the slackened upper wishbone arm re-positioned and tightened to the shock absorber arms.

Note that the swivel pin and upper trunnion fulcrum pin nuts must not be tightened at this stage.

Replacing the Suspension

Fit one rubber bearing to each of the suspension lower links, on the side which corresponds to the small hole in each of the frame brackets.

Raise the links to the frame brackets, insert the fulcrum pins and slide the second bearing and special washer over the protruding end of each pin. Fit the nut but do not screw it home. Position the shock absorber on its top bracket and partially tighten the four setscrews.

The assembly must next be set in the normal loaded position. This can be accomplished by placing a distance piece between the shock absorber wishbone arm and the upper spring plate at a point opposite the rubber buffer. The length of the distance piece must be 2 in. The final

adjustments can now be effected as follows :—

1. Tighten the nuts on the fulcrum pins securing the lower wishbone arms to the frame brackets. Do not forget to lock them with the split pins.

2. Tighten the four setscrews securing the shock absorber to its bracket on the frame.

3. Tighten the upper trunnion fulcrum pin nut and secure with a split pin.

4. Tighten the swivel pin nut and lock with a split pin.

Service Tool 18G 56, or the lower spring plate, whichever used, should now be removed from the lower wishbone arms and the coil spring refitted as already described.

Connect the brake fluid pipe to the brake backplate, secure the steering side-tube to the steering arm, refit the road wheel, lower the car to the ground and remove the distance piece used to retain the suspension in the normal loaded position.

Finally, bleed the brakes as described in section **M**.

AH. 184. A.

Fig. 10.   When building up the suspension, the arms must be correctly set by the distance piece A (2 in.) before the various bearings are tightened.

CASTOR AND CAMBER ANGLES AND SWIVEL PIN INCLINATION

Description

The castor and camber angles and the swivel pin inclination are three design settings of the front suspension assembly. They have a very important bearing on the steering and general riding of the car. Each of these settings is determined by the machining and assembly of the component parts during manufacture. They are not therefore adjustable.

However, should the car suffer damage to the sus-

pension affecting these settings, the various angles must be verified to ensure whether replacements are necessary.

Camber Angle

This is the outward tilt of the wheel and a rough check can be made by measuring the distance from the outside wall of the tyre, immediately below the hub, to a plumb line hanging from the outside wall of the tyre above the hub. The distance must be the same on both

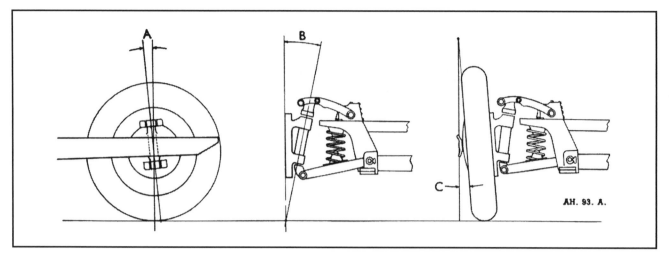

Fig. 11.

A. Castor angle 1¼°.   *B. Swivel pin inclination,* 6½°.   *C. Camber angle* 1°.

wheels. Before making this test, it is very important to ensure that the tyres are in a uniform condition and at the same pressure. Also that the car is unladen and on level ground.

Damage to the upper and lower wishbone arms may well affect the camber angle.

Castor Angle

This is the tilt of the swivel pin when viewed from the side of the car. This also is only likely to be affected by damage to the upper and lower wishbone arms.

Swivel Pin Inclination

This is the tilt of the swivel pin when viewed from the front of the car and is again only likely to be affected by damage to the wishbone arms.

A useful tool which can be used for checking these settings is the Dunlop "wheel camber, castor and swivel pin gauge". With the car standing on level ground this gauge will give readings enabling the castor, camber and swivel pin angles to be quickly verified. Full details of this gauge can be obtained from the Dunlop Rubber Co. Ltd., Fort Dunlop, Erdington, Birmingham.

STEERING

GENERAL DATA

Type of Gear	Cam and Lever
Maker's Number	L 3 A
Steering Gear Ratio	12.6 to 1
Bearings...	Ball Race and Felt Bush
Adjustment	Packing Shims
Diameter of Steering Wheel...	16½ ins. (41.91 cm.)
Turning Circle...	...35 ft. (10.668 m.)
Track Toe-in	$\frac{1}{16}$ to $\frac{1}{8}$ in. (1.588–3.175 mm.)
Steering Connections	Ball and Socket Type

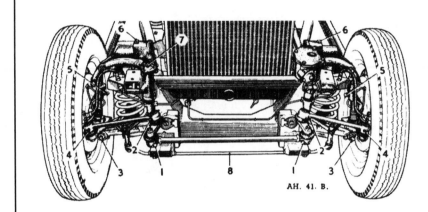

AH. 41. B.

Fig. 1. Showing the front suspension and steering layout and lubrication points.

1. Cross tube connections.
2. Side rod inner connections.
3. Lower link.
4. Side rod outer connections.
5. Swivel pin.
6. Shock absorber.
7. Steering idler.
8. Anti-roll bar (no lubrication required).

Description

The "Burman" Steering gear is a self-contained unit consisting of a single start worm at the lower end of the column, supported by two ball races. The gear is housed within an oil tight steering box, mounted in front and to one side of the radiator. An oil impregnated felt bush supports the column at its upper end.

A spring loaded peg, mounted in two ball races in the rocker shaft, engages the worm.

Turning the steering column produces a rotating action of the follower peg and an arcuate movement of the rocker shaft. As the peg moves out of the plane of the worm the rocker shaft is moved inwards, thus keeping the peg in contact with the worm by means of a cam, in turn controlled by a fixed thrust button in the top cover plate.

Attached to the rocker shaft is the steering lever which is connected to the steering linkage.

The steering linkage consists of a cross-tube which is connected to the steering lever on one side and the idler lever on the other. Two side rods, one at each side, connect the swivel arms to the steering and idler levers respectively.

Maintenance

Lubrication of the oil nipples on the steering connections and swivel bearings is most important to maintain accurate steering.

Approximately every 1,000 miles, use the oil gun and recommended oil to charge the following points with lubricant :—

(a) Steering side rods and cross tube—6 nipples.
(b) Lower wishbone arm outer housing—2 nipples.
(c) Swivel pin bushes—4 nipples.
(d) Steering idler—oil filler plug.

The steering box should be topped up with recommended oil to the top of the filler plug opening approximately every 1,000 miles.

STEERING CONNECTIONS AND IDLER

Adjusting the track

The track is best adjusted by means of a Dunlop Optical Alignment Gauge, particulars of which can be obtained from the Dunlop Rubber Co. Ltd., Fort Dunlop, Erdington, Birmingham.

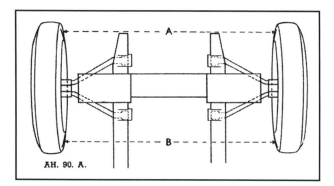

Fig. 2. The toe-in must be adjusted so that A is 1/16 to 1/8 inch less than B.

The cross-tube is threaded right-hand at one end and left-hand at the other, so that the track adjustment can be made by simply rotating the tube in the required direction after releasing the locknuts. Always re-tighten the locknuts at each end of the cross-tube after an adjustment has been made.

The side-rods are non-adjustable.

When adjusting the track the following precautions should be observed:—

(1) The car should have come to rest from a forward movement. This ensures as far as possible that the wheels are in their natural running position.

(2) It is preferable for alignment to be checked with car laden.

With conventional base-bar tyre alignment gauges

(3) measurements in front of and behind the wheel

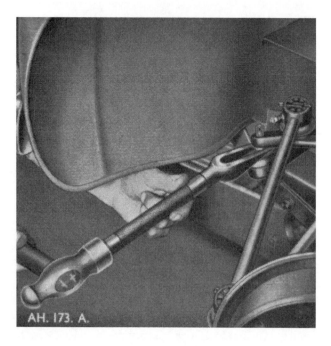

Fig. 3 Using Service Tool 18G 125 to separate the ball and socket connection from the steering lever.

centres should be taken at the same points on the tyres or rim flanges. This is achieved by marking the tyres where the first reading is taken and moving the car forwards approximately half a road wheel revolution before taking the second reading at the same points. With the Dunlop Optical Gauge two or three readings should be taken with the car moved forwards to different positions— 180° road wheel turn for two readings and 120° for three readings. An average figure should then be calculated.

Wheels and tyres vary laterally within their manufacturing tolerances, or as the result of service, and alignment figures obtained without moving the car are unreliable.

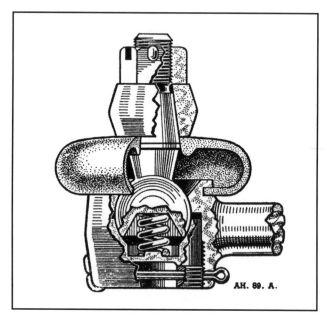

Fig. 4. A cutaway view of the adjustable ball and socket connection.

Cross Tube and Side Rods

The cross-tube and side rods are held in position by a castellated nut and split pin at each end. On early models a plain washer is fitted beneath the castellated nut. It is essential that this washer be replaced on re-assembly, otherwise the nut, when tightened, will be screwed too far down the thread to enable the split pin to lock the nut in position.

To remove either the cross-tube or a side rod, withdraw the split pin and release the nut at each end. Using Service Tool 18G 125 (see fig. 3), separate the ball and socket connections from the levers in the following manner. Insert the tool between the connections and the lever so that the inclined face of the tool is against the lever. One sharp blow with a hammer will usually be sufficient to separate the components.

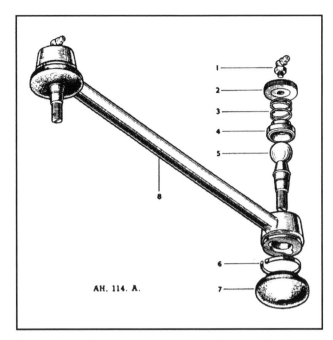

Fig. 5. An exploded view of a non-adjustable type ball and socket connection.
1. *Oil nipple.* 2. *Screwed fixing plate.* 3. *Spring.* 4. *Socket.* 5. *Ball.* 6. *Spring clip.* 7. *Rubber boot.* 8. *Side rod.*

Steering Connections

On early models the cross-tube and side rods are equipped with adjustable type ball and socket connections, see fig. 4. These connections consist of a threaded and castellated lower socket screwed into position and locked by a split pin. The body of the tube end has four split pin holes drilled, vernier pattern, at a different pitch from the castellations in the socket, thus permitting a very fine adjustment. **Adjustments should be made and checked regularly,** otherwise undue slackness will cause a deformity of the ball pin thereby making further adjustment impossible.

To make an adjustment, remove the split pin, lightly screw up the socket as far as it will go, then screw it back to the first alignment of the split pin hole and castellation. The ball should then be able to move freely in the socket. See that the rubber boot fits snugly in its groove.

Later models are fitted with a non-adjustable type of ball and socket connection, differing slightly in construction to the adjustable type. See fig. 5. Both types should be lubricated at the prescribed intervals.

Steering Levers

These are held to the steering gear rocker shaft and to the idler shaft respectively by a castellated nut, plain washer and split pin.

Only under rare circumstances does the steering lever have to be removed from the steering gear rocker shaft splines, such as suspected damage to the lever, but not to the steering box or column.

In such circumstances the steering column must be released, moved bodily downwards and then turned through 90° in order to use the steering lever extractor 18G 75 (see fig. 6). To disconnect the column proceed as for first five paragraphs of Steering Column Removal, page K/5.

More generally the lever need not be removed from the steering gear rocker shaft unless the complete steering gear has to be dismantled. In this instance the lever can be left in place until the steering column has been removed from the car.

The steering idler lever must only be removed (using 18G 75) when the idler has been released from the chassis. See "Removing the Idler".

The levers must be refitted with the location marks facing downwards. In the case of the steering gear lever, this mark must be lined up with the corresponding mark on the steering gear rocker shaft. Do not forget to replace the sponge rubber sealing rings.

Removing the Idler

With the side rod and cross-tube disconnected the idler, together with its lever, can be detached from the chassis. It is held by three setpins which pass through a bracket, incorporating an aluminium distance piece, and terminating in the three tapped holes in the idler flange. Unscrew the setpins until their threads are clear of the holes and remove the idler.

Dismantling the Idler

The idler top cap is secured to the body by three setscrews and has a joint washer inserted between cap and body. Lubrication is effected by removing the screwed plug in the centre of the cap and injecting oil into the body.

Fig. 6. Using Service Tool 18G 75 to remove steering lever.

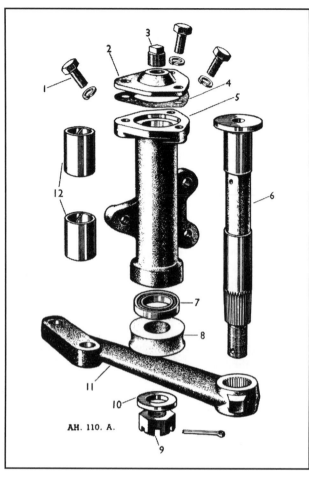

Fig. 7. Steering idler exploded.

1. Cap setpin.	7. Oil seal.
2. Idler cap.	8. Dust excluder.
3. Oil plug.	9. Castellated nut.
4. Joint washer.	10. Plain washer.
5. Idler body	11. Steering lever.
6. Idler shaft.	12. Bush bearings.

The idler shaft can be removed by hand once the body cap has been released. The flange of the shaft locates in the recess within the body head and the two highly finished portions of the shaft rotate within the phosphor bronze bushes. At the lower end of the body there is a recess to take the rubber oil seal. The idler shaft is drilled for passing lubricant to the bearing bushes.

Swivel Arms

The swivel arms, connecting the swivel axles to the side rods, may be checked for misalignment in the following way :—

(1) Place a rule along the brake backplate so that it projects along the arm. The distance between the centre of the ball pin hole and the rule should be $\frac{7}{8}$ in. plus or minus $\frac{1}{16}$ in.

(2) Place a straight-edge across the centre of the bolt holes used to secure the arm to the swivel axle. The distance between the straight-edge and the

lower face of the arm—machined face against which the ball pin fits—should be $1\frac{1}{16}$ in. plus or minus $\frac{1}{16}$ in. (see fig. 8).

STEERING GEAR

Steering Wheel Removal

Early Models : The steering wheel fitted to the early models is of the adjustable type. For this reason the stator tube is made in two parts; a short piece of tube, attached to the horn quadrant, fits over the end of the main stator tube. The main tube is secured at the steering box end of the column by a nut and olive type gland washer.

To release the steering wheel, first disconnect, at the nearest snap connections, the horn and flasher light cables protruding from the end of the stator tube. Slacken the two grub screws in the steering wheel hub and pull out the horn quadrant together with the short piece of stator tube, horn and flasher light cables. It will be noticed that the short tube has a number of indentions on its outer diameter, thus forming lugs internally, which locate in the slot in the long tube remaining within the steering column.

Prise off the circlip exposed to view, then release the locking ring situated behind the steering wheel hub. The

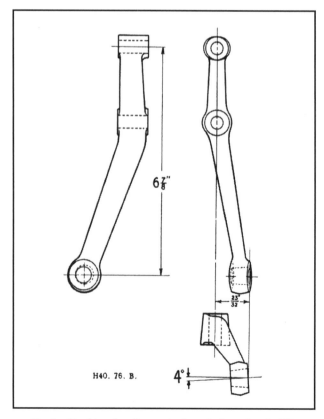

Fig. 8. Checking the alignment of a swivel arm (method 2).

steering wheel can now be pulled clear of the column, followed by the telescopic spring and locating collar.

Later Models : These models have a non-adjustable steering wheel and a one piece stator tube, secured at its lower end in a similar manner to that previously described.

Fig. 9. Using Service Tool 18G 70 to remove a tight steering wheel.

At the steering box end of the column, the horn and flasher light cables should be disconnected at their nearest snap connections. Release the nut and olive type gland washer securing the stator tube at the steering box end of the column and withdraw the quadrant assembly from the column.

Using Service Tool 18G 38 or a box spanner, undo the steering wheel nut and take off the serrated washer and trip lever. A sharp jerk should be sufficient to release the steering wheel, but should the wheel prove to be tight on the splines, use extractor 18G 70 (fig. 9).

Steering Column Removal

With the stator tube and steering wheel removed, the steering column removal procedure is the same for both early and later models.

From behind the instrument panel release the two-piece clamping bracket supporting the top end of the column. The clamp is held by two bolts with nuts and washers; the shorter of the two bolts grips the lower ends of the clamp halves. When refitting, do not forget to replace the strip of rubber between clamp and column.

There are two sealing plates, one either side of the scuttle, through which the steering column passes. Release each plate by undoing four metal thread screws, thus giving more room when finally manœuvring the column from the car. Remember, when refitting the column, to replace the rubber grommet and large circular felt washer with each plate.

Jack up the front of the car and remove the front wheels. Now disconnect the cross tube and side rod from the steering lever as previously described.

The steering gear box is secured to the chassis by three bolts with nuts and washers, which should now be released. The bolts pass through a bracket incorporating an aluminium distance piece which must be replaced before attempting to secure the steering gear box in position. If it is desired at this stage to remove the steering lever, the steering column should be pulled forward and turned through ninety degrees to enable the steering lever to be extracted. See fig. 6.

Undo the radiator fixing points (see fig. 8, page B/4). There is no need to drain the cooling system or disconnect the water hoses. Just move the radiator a little to one side and this will give sufficient room to manœuvre the column out through the radiator grille aperture.

Fig. 10. Showing the steering column being manœuvred out through the radiator grille aperture.

When replacing the steering column, reverse the removal procedure, but do not tighten the steering box nuts and bolts until the column has been secured within the driving compartment. Refit the steering wheel so that one of the spokes points vertically upwards; the front wheels being in the straight ahead position.

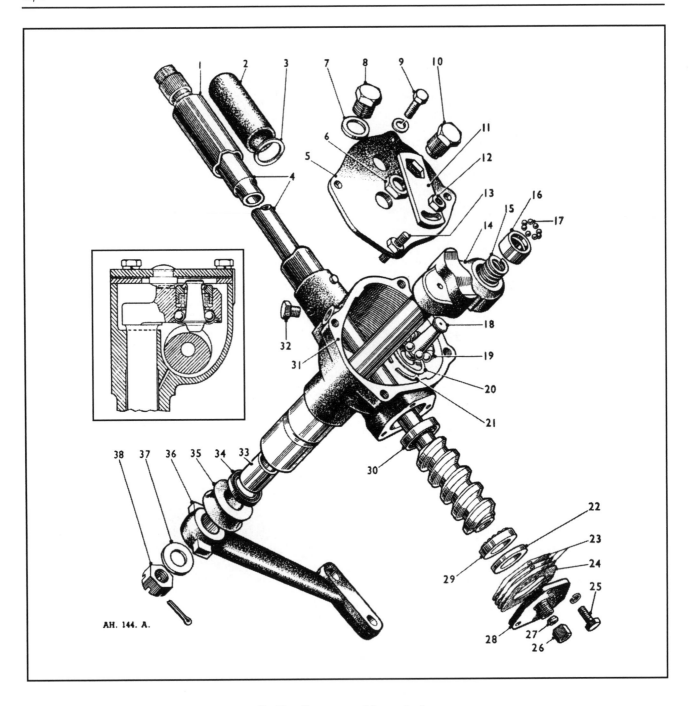

Fig. 11.   *Components of the steering box.*

1. *Outer column.*	13. *Locking plate setpin.*	26. *Stator tube nut.*
2. *Felt bush.*	14. *Follower peg housing.*	27. *Olive type gland washer.*
3. *Washer.*	15. *Spring.*	28. *End cover.*
4. *Inner column.*	16. *Ball race housing.*	29. *Column lower ball race.*
5. *Top cover.*	17. *Ball bearings.*	30. *Column upper ball race.*
6. *Locknut.*	18. *Follower peg.*	31. *Steering box.*
7. *Washer.*	19. *Ball bearings.*	32. *Oil filler.*
8. *Alternative filler plug.*	20. *Dished plate.*	33. *Bush bearing.*
(Later models only.)	21. *Circlip.*	34. *Oil seal.*
9. *Top cover setpin.*	22. *Distance piece.*	35. *Dust excluder.*
10. *Adjusting screw.*	23. *Adjusting shims.*	36. *Steering lever.*
11. *Locking plate.*	24. *Joint washer.*	37. *Washer.*
12. *Locking plate nut.*	25. *End cover setpin.*	38. *Castellated nut.*

Inset: Cross sectional view of peg, cam and top plate with
non-adjustable thrust button, as fitted to early models.

Dismantling the Steering Gear

Early Models: If the steering lever has not already been removed, release the castellated nut and split pin, then use Service Tool 18G 75 to extract the lever.

The top cover plate, together with the adjustment shims, should be removed after extracting the four securing setpins. The rocker shaft and the follower peg can now be easily removed by hand. The peg is spring loaded and mounted on two ball races. To release the peg, it is first necessary to press down on the upper ball race housing (which is also the spring housing) to release the ball bearings. The lower race, which is the larger of the two, is held in place by a dished plate and circlip; with these removed the follower peg and ball bearings will come away in the hand. The assembly is shown in exploded form in fig. 11.

Before attempting to remove the inner column, release the stator tube securing nut and washer at the steering box end of the column and withdraw the tube.

Release the three setpins holding the end cover plate in place. Remove the cover plate, but be careful not to lose or damage the adjustment shims. The complete unit should now be up-ended with the steering box uppermost. By bumping the end of the inner shaft on a block of wood, placed on the floor, the worm with its two ball races will be displaced. The complete inner column can then be withdrawn from the casing through the open end of the steering box.

For extracting the felt bush at the top of the column, use a piece of strong hooked wire, the hook pulling on the under face of the bush. The fitting of a new bush is quite simple; smear the felt bush with heavy oil and press into place.

Adjustment shims should be fitted behind the end cover so that there is no play on the column, but at the same time the bearings should not be preloaded otherwise they may be damaged.

Having replaced the follower peg, the rocker shaft may now be dropped into position. Ensure that the rocker shaft is a good fit in its housing and the oil seal at the lower end of the housing is making good contact with the shaft.

Later Models: With the exception of the adjusting screw locking plate (see fig. 11), which should be removed before the top cover is released, the procedure just described should be adopted. Remember that the stator tube will have already been withdrawn.

Adjusting the Gear

Early Models: The method of adjustment of the rocker shaft peg and cam engagement is effected by placing a sufficient number of shims between the steering box and cover plate to take up the free movement of the gear in the straight ahead position. The cover plate incorporates a fixed thrust button.

A final check of the adjustment should be made when the gear has been refitted to the chassis. It should be noted that as wear in use is normally greater in the straight ahead position than on lock, provision is made for this in the design of the cam, and it will be found that there is slight end play towards each lock. It is essential, therefore, that adjustment should be made in the straight ahead position to avoid the possibility of tightness.

The steering box should be filled with recommended oil and a test made to ensure that the movement is free from lock to lock.

Later Models: The adjustment on these models is effected by an adjusting screw in the cover plate.

The procedure to be adopted for this type of adjustment is first to slacken the adjusting screw locknut, then unscrew the locking plate nut an amount sufficient to allow the plate to swing clear of the adjusting screw head. Screw down the adjusting screw until there is no free movement of the gear in the straight ahead position. Refit the locking plate over the adjusting screw and secure with the locking plate nut. Finally, retighten the locknut.

The object of the locking plate is to prevent the adjusting screw from moving when the locknut is tightened. The plate has been designed so that it will fit over the head of the adjusting screw no matter what position it should occupy.

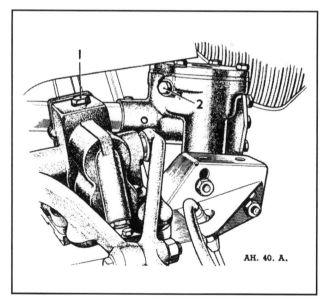

Fig. 12. Showing at 1 the shock absorber filler plug and at 2 the steering box filler plug.

A final test should be carried out when the steering gear is installed, and, as previously described, a check made for free movement when filled with oil.

Once again it is stressed that the adjustment must only be made in the straight ahead position.

STEERING FAULTS

If steering faults are not attributed to adjustment of the gear, they may fall into one of the following categories.

Lost Motion

The amount of lost motion reaches its maximum at either lock, but this is not normally felt at the steering wheel, as the geometry of the steering always tends to return the steering gear to the straight ahead position.

Excessive lost motion in the steering gear will result in unsteady steering, knocks and backlash all of which can be felt in the steering wheel. This defect may be attributed to loose steering connections throughout the linkage.

Tight Steering

If the steering is tight, disconnect the steering tubes and test the feel of the steering wheel. Stiffness may be due to the steering column being pulled out of line and this can be verified by loosening the column supporting brackets under the fascia, and allowing the column to find its free position. Should the steering still be stiff check whether this is so in all positions. If so, the cause may be :—

(*a*) The stator tube is fouling the column.

(*b*) The felt bush at the top of the steering column is too tight.

(*c*) The steering tube is bent.

To ascertain whether the stator tube is fouling the column, withdraw the tube as previously described. Turn the steering wheel and if the stiffness has disappeared, it will probably be found that the stator tube is bent, thus requiring a replacement.

Should the steering still be stiff with the stator tube free, withdraw the steering wheel and check the tightness of the felt bush and renew if necessary.

If the bush is free but the steering remains tight, remove the bush and check whether the inner steering column pulls heavily to one side. The inner column is fairly flexible and slight pulling to one side has little or no effect on the feel of the gear, but it may be that the column is bent thus giving no alternative but to renew the column.

Loose Steering

Loose steering is invariably attributed to end play of the inner column, which can be rectified by the removal of shims located behind the steering box end cover plate, in a manner already described.

To check for this end float, disconnect the side and cross-tubes from the steering lever and turn the steering partly to the left or right lock. Then with the steering wheel held to prevent it from turning, endeavour to turn the steering lever. Should the steering wheel have a tendency to lift, it may be assumed that there is end float of the gear.

SERVICE DIAGNOSIS GUIDE

Symptom	No.	Possible Cause
(a) Wheel Wobble	1	Unbalanced wheels and tyres
	2	Slack steering connections
	3	Incorrect steering geometry
	4	Excessive play in steering gear
	5	Broken or weak front springs
	6	Loose idler mounting or worn idler shaft
	7	Worn hub bearings
	8	Loose or broken shackles
(b) Wander		Check 2, 3, 4 and 8 in (a)
	1	Broken spring clips
	2	Front suspension and rear axle mounting points out of alignment
	3	Uneven tyre pressures
	4	Uneven tyre wear
	5	Weak shock absorbers or springs
(c) Heavy Steering		Check 3 in (a)
	1	Excessively low tyre pressures
	2	Insufficient lubricant in steering box
	3	Insufficient idler lubrication
	4	'Dry' steering connections
	5	Out of track
	6	Incorrectly adjusted steering gear
	7	Misaligned steering column
(d) Tyre Squeal		Check 3 in (a)
		Check 1 in (c)

SERVICE JOURNAL REFERENCE

NUMBER	DATE	SUBJECT	CHANGES

REAR AXLE AND SUSPENSION

AXLE DATA

Type ...	¾ Floating
Oil Capacity ...	2¼ Imp. pints 2.7 US. pints, 1.28 litres
Final Drive ...	Spiral Bevel
Crown Wheel Teeth ...	33
Bevel Pinion Teeth ...	8
Ratio ...	4.125 to 1
Crown Wheel to Pinion Backlash ...	.005–.008 in.

Bearings :—
Pinion Front

Make ...	Timken/Skefko
Type ...	Taper Roller
Size...	$1 \times 2.5 \times \frac{5}{8}$ in. (25.4×63.5×15.875 mm.)

Pinion Rear

Make ...	Timken/Skefko
Type ...	Taper Roller
Size...	1.25×2.86×.94 in. (26.03×72.62×23.81 mm.)

Differential

Make ...	R & M. LJT.40
Type ...	Double Purpose Ball Journal (light)
Size...	40×80×18 mm.

Hub

Make ...	R & M. LOJ.40
Type ...	Ball Journal Double Row (light)
Size...	40×80×23 mm.

AXLE UNIT REMOVAL

Loosen the hub caps, then jack-up the car and place supports under frame members just forward of rear springs front anchorage. Take off both wheels after removing the hub caps. Working from under the car, knock back the lockwasher tabs and remove the four bolts and nuts securing propeller shaft flange to axle pinion flange.

The handbrake cable should next be disconnected from the axle. This is accomplished by unscrewing it from its link to the brake balance lever, and unscrewing the nut holding its outer casing to the axle. The hydraulic brake pipe at the rear axle is detached from the flexible pipe at the union just forward of the right-hand shock absorber.

Pull out the split pins from the castellated nuts securing the shock absorber links to the axle mounting brackets. Unscrew the nuts completely but do not attempt to remove the links as this operation will prove much easier when freeing the axle.

The right-hand rebound rubbers, one on the chassis and one above the spring fixed to the boot floor, should be completely removed for satisfactory axle extraction. Two setpins fix the chassis rebound rubber in place.

Note :—On refitting, the tapered end of the rubber faces forwards. The larger rebound rubber is best detached by peeling back the required amount of lining on the rear floor of the boot to give access to the four bolt heads.

Remove the negative terminal from the battery tops and unscrew the two large bolts positioned to hold each battery. The batteries may then be removed through the opening at the rear of the seats.

Next, remove the Simmonds self-locking nuts from the spring clips ("U" bolts) which secure the axle to the springs. Observe that a fibre pad and aluminium distance piece are situated between the axle and spring. The tapered end of the aluminium piece faces the rear of the car. Disconnection of the left-hand rear "U" bolt releases the anti-sway bar.

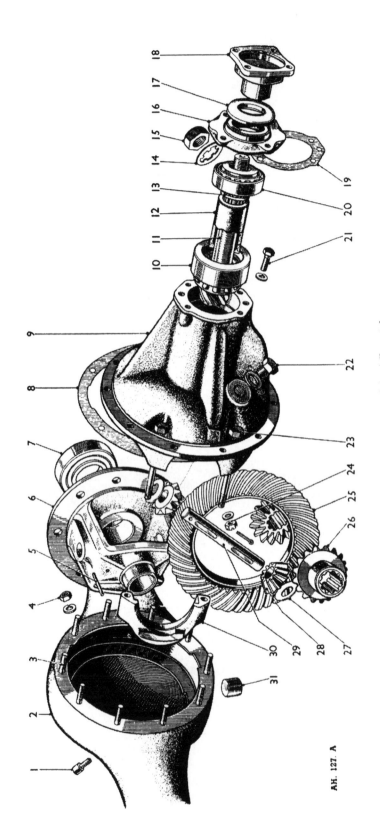

Fig. 1. An exploded view of the differential.

AH. 127. A

1. Axle breather.
2. Axle case.
3. Gear carrier stud.
4. Gear carrier nut.
5. Pinion shaft securing pin.
6. Differential case.
7. Crown wheel bearing.
8. Joint washer.

9. Gear carrier.
10. Bevel pinion rear bearing.
11. Bevel pinion.
12. Pinion sleeve.
13. Bearing adjusting shim.
14. Bevel pinion lockwasher.
15. Bevel pinion nut.
16. End cover.

17. Dust cover.
18. Pinion flange.
19. Joint washer.
20. Bevel pinion front bearing.
21. End cover bolt.
22. Oil filler plug.
23. Crown wheel bearing cap bolt.
24. Nut for bearing cap bolt.

25. Crown wheel.
26. Differential wheel.
27. Pinion thrust washer.
28. Differential pinion.
29. Differential pinion shaft.
30. Crown wheel bearing cap.
31. Axle drain plug.

Note:—The bevel pin can only be driven out of the carrier towards the crown wheel.

With the axle free to move, the connecting links from the shock absorbers should be detached.

The complete axle should be removed, for further dismantling, from the left-hand side of the car. Take every care not to damage other components, particularly the petrol pump.

Installing the axle is the reverse of the above operations.

On re-assembling it is advisable to jack-up the springs to meet the axle thus locating the spring centre bolt properly and easing the fixing of the "U" bolts. Remember to fit the fibre washer and the aluminium distance piece which fits over it.

When assembly is complete adjust the handbrake if required and bleed the hydraulic brake system all round.

MAINTENANCE

Whilst the following maintenance dismantling routines are described as for the axle being left in position on the car, the operator may remove the axle for his own convenience to work on such items as the gear carrier and bevel pinions on a bench.

Lubrication

For the lubrication of the axle use lubricants from approved sources only as tabulated on page S/2. To change axle oil, remove the plug at the base of the centre casing, drain away the old oil, replace the plug, then remove the filler plug, see fig. 2, and top-up with 2¼ pints of new oil.

Axle Shaft—To Remove and Replace

Loosen the hub cap of the wheel concerned before jacking-up the car. Remove the wheel after further unscrewing the hub cap, thus giving access to the rear hub extension. This is secured by four self-locking nuts.

Next, take out the drum locating screw, using a screwdriver. The drum can be tapped off the hub and brake linings, provided the handbrake is released and the brake shoes are not adjusted so closely as to bind on the drum.

Should the brake linings hold the drum when the handbrake is released, it will be found necessary to slacken the brake shoe adjuster a few notches.

Remove the axle shaft retaining screw and draw out the axle shaft by gripping the flange outside the hub. It should slide easily but if it is tight on the studs it may need gently prising with a screwdriver inserted between the flange and the hub. Should the paper washer be damaged it must be renewed when re-assembling.

Replacement is a reversal of the above operations.

Hubs—To Withdraw and Replace

Remove the wheel and axle shaft as described, when the hub retaining nut will be accessible. This nut is

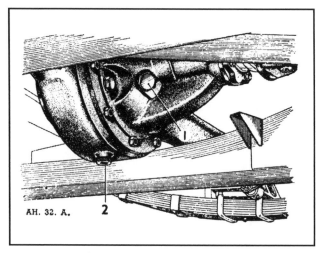

Fig. 2. *A view of the axle from beneath the car to show* 1. *Filler plug.* 2. *Drain plug.*

locked in position by a keyed washer which is hammered down on to one of the flats of the nut. Knock back the washer and remove the nut with a well fitting spanner.

The lockwasher can be removed by hand by tilting it so that the key disengages with the slot in the threaded portion of the axle case.

To use the extractor on the rear hub the adaptor will be needed. It will be seen that this piece fits into the end of the axle tube and provides a stop for the extractor bolt.

The extractor is fitted over the wheel studs and held in position by two wheel nuts screwed well down. By screwing up the central bolt of the extractor, using either a spanner or a tommy bar, the hub and double-row ball bearing, together with washers and oil seal will be withdrawn. Fig. 4 shows the assembly order.

Fig. 3. *Using an extractor to withdraw a rear hub.*

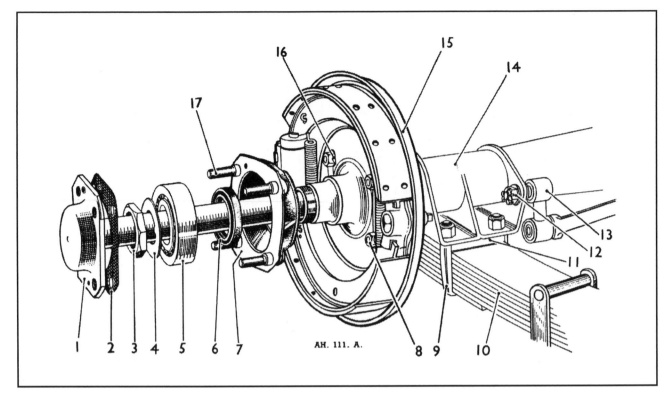

Fig. 4. Rear axle hub assembly in exploded form.

1. *Axle shaft.*
2. *Joint washer.*
3. *Hub locknut.*
4. *Hub lockwasher.*
5. *Hub bearing.*
6. *Oil seal.*

7. *Hub casing.*
8. *Backplate nut.*
9. *Spring "U" bolt.*
10. *Spring.*
11. *Packing plate.*
12. *Link nut.*

13. *Shock absorber link.*
14. *Axle tube.*
15. *Backplate.*
16. *Backplate nut lockwasher.*
17. *Hub extension stud.*

The bearing can be tapped out of the hub with the aid of a drift.

To Re-assemble

The hub bearing is not adjustable and is replaced in one operation.

It is essential that the face of the outer race protrudes .001 in. to .004 in. beyond the face of the hub plus paper washer when the bearing is pressed into place. This ensures that the bearing is definitely gripped between the abutment shoulder in the hub and the flange of the differential shaft.

The hub is then mounted on the axle tube, followed by the lockwasher (which has a tongue to register with the groove or hole) and finally the securing nut.

Tighten up the nut until the hub is fully home, and then secure by tapping down the lockwasher on one of the flats of the nut.

Replace the axle shaft, carefully finding the spline engagement with the differential unit and ensuring that the flange and washer are threaded over the hub studs in the position in which the two small holes of the flange and hub coincide. Then refit the brake drum taking the

same precautions regarding the two small holes, and ensuring that the drum is well home when inserting the screws. Temporary use of the wheel nuts will assist.

Replace the wheel and finally tighten the nuts.

During re-assembly the hubs should be packed with fresh grease even though they receive some lubricant from the axle during normal running.

Bevel Pinion—To Renew the Oil Seal

Knock back the lockwashers and take out the four bolts of the propeller shaft flange to axle pinion flange.

Remove the large nut in the centre of the pinion flange after knocking back the lockwasher and then withdraw the flange itself. A flange extractor should be used, such as Service Tool 18G 2, but it may be possible to tap the flange off the splined pinion shaft.

Remove the four setpins from the pinion end cover after knocking back the tab washers, when the end cover can be withdrawn.

The oil seal is pressed into this end cover, but can be removed with a punch. The end cover is of aluminium and care must be used to prevent damage. Never remove an oil seal from the end cover unless it is intended to renew it, as it is invariably distorted in removal.

The new oil seal must be carefully pressed home with the edge of the rubber or leather sealing ring facing inwards.

Replace the end cover and paper washer and lock up the four setpins when thoroughly tightened. Replace the pinion flange, serrated washer, and nut. This nut must be fully tightened and finally locked in position by bending over the lip of the washer. Use new lock-washers when bolting up the propeller shaft.

The Carrier—To Withdraw

To renew the pinion bearings or to effect any servicing of the crown wheel and differential unit it is first necessary to withdraw the gear carrier.

The gear carrier unit can be withdrawn with the axle in position, although it is first necessary to remove the propeller shaft and then the axle shafts. For this latter operation the road wheels and brake drums must also be removed as already described under "Axle shaft removal". Both batteries should be removed as described under "Axle Unit Removal".

Remove the gear carrier unit drain plug and run the oil into a suitable receptacle.

Then remove the nuts which hold the gear carrier to the axle case and lift out the carrier complete while holding propeller shaft as high as possible in tunnel.

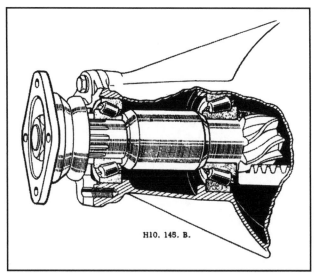

H10. 146. B.

Fig. 6. The taper roller bearings are subject to a pre-load of 6-inch lbs. controlled by shims fitted between the pinion sleeve and the inner races of the bearings.

Pinion and Bearings

The pinion has two taper roller bearings, the larger in the rear, and for removal the shaft has to be driven out rearwards from the propeller shaft end after the crown wheel has been removed by releasing the main bearing caps and lifting off the crown wheel complete with the differential carrier. The pinion will bring with

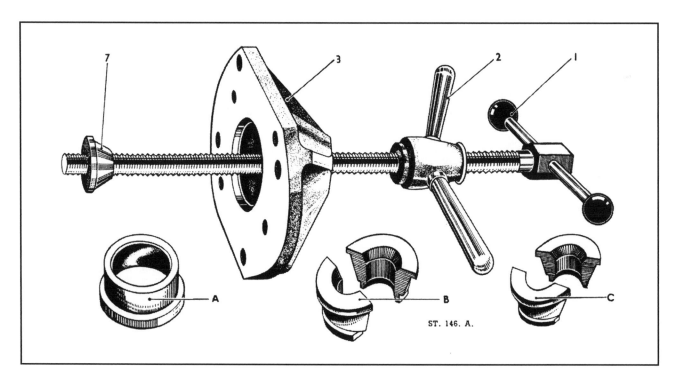

ST. 146. A.

Fig. 5. Service Tool 18G 82 to remove pinion and outer races.
1. Tommy bar.   2. Wing nut.   3. Tool body.   7. Cone.
A. Adapter ring.   B. Split adapter for removing and replacing rear outer race.   C. Split adapter for removing and replacing front outer race.

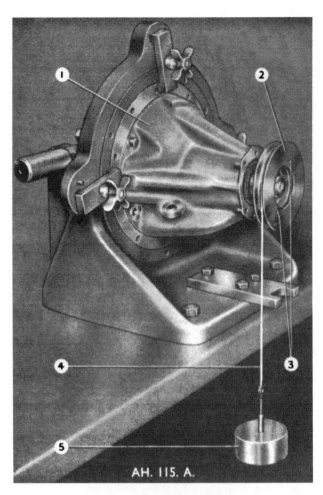

Fig. 7. Checking pinion pre-load.
1. *Gear carrier.* 2. *Pulley.* 3. *Riveted ends of locating pegs.* 4. *Cord.*
5. *Weight.*

it the inner race and rollers of the rear bearing, and there will be left in the case the rear outer race and the front roller bearing complete.

The inner race of the front bearing can be removed with the fingers. The outer races can be tapped out with a soft drift or suitable extractor, such as Service Tool 18G 82, fig. 5.

Replacement of the pinion races must be carefully carried out to ensure that they are fully home against the shoulder in each case, and perfectly square with the housing. First fit the outer races to the housing, then the inner race and rollers of the rear bearing to the pinion, slide on the pinion sleeve and shim or shims as found on dismantling, and push into position from the rear. Fit the inner race and rollers of the front bearing over the splined end of the pinion and tap into position. Replace oil seal housing and paper washer, dust cover, propeller shaft flange, serrated washer and nut. Tighten the nut well home.

The taper roller bearing assembly is designed for pre-loading to the extent of 6 to 8-inch lbs. drag excluding drag of the oil seal.

A convenient method of checking this pre-loading is by mounting a 4-inch diameter drum or pulley to the pinion propeller shaft flange. Then when a 3-lb. weight is hung by a piece of cord from the circumference of the pulley, the pre-loading on the pinion bearings should be just sufficient to prevent the pinion, and therefore the pulley, from turning. When a 5-lb. weight is used the pulley should move under the load. This gives the bearings a pre-load of approximately 6 to 8-inch lbs.

Adjustment is effected by the insertion or removal of shims between the pinion sleeve and the inner races of the pinion bearings as necessary to reduce or increase the degree of pre-loading.

Crown Wheel and Bearing Removal

Disengage the gear carrier as detailed above; then remove the bearing caps (drift them off) and lever out the differential case and bearings (fig. 8). Undo the lockplates and bolts, then take off the crown wheel.

The ball races should be a tight fit on each end of the differential case and if found to be loose, a new case will be needed. A bearing extractor should be used to remove them. Preserve the shims under each bearing for re-use.

Removal of Differential Wheels and Pinions

The differential case is of one-piece construction and cannot be dismantled.

To release the differential wheels and pinions tap out the peg securing the pinion shaft. Extract the shaft and remove the differential wheels, pinions and thrust washers.

As the unit is now fully dismantled all parts should be checked for scoring and signs of wear.

Wash all the components in paraffin and ensure that they are clean when re-assembled.

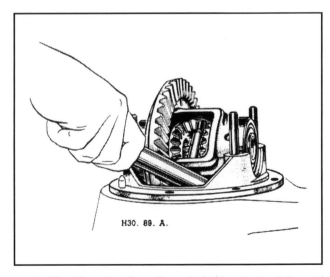

Fig. 8. This illustration shows the method of levering the differential unit out of the carrier.

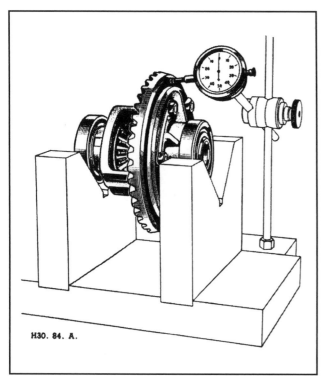

H30. 84. A.

Fig. 9. Checking crown wheel run-out with a dial indicator gauge.

Assembling Differential Wheels and Pinions

Replace the differential wheels in the case and place the pinions with their thrust washers, in position. Push home the pinion shaft and line the hole up to allow the peg to be driven in. Fit the peg through the hole provided and pein the rim over to prevent the peg vibrating loose.

Replacing Crown Wheel and Bearings

Bolt the crown wheel to the case, tightening alternately and evenly, but do not turn over the lockwashers. If new races are necessary take particular care in ensuring that they are correctly fitted using a suitable hand press. the word "Thrust" which is stamped on one side of the race must be on the outside.

At this stage the crown wheel should be checked for alignment (fig. 9) by placing the crown wheel, differential case and bearings complete on a pair of vee blocks and using a dial indicator. As the bearings are resting in the vee blocks, the differential case can be rotated by turning the crown wheel and readings taken off the face of the wheel. The crown wheel must not be more than .002 in. (.0508 mm.) out of true. Any greater irregularity must be corrected.

First detach the crown wheel from the differential casing and look for small particles of grit on the flanges.

It will be found that the crown wheel will not often run out of true if the two surfaces have been thoroughly cleaned. The pinion mesh will be automatically correct for depth.

Finally, turn over the lockwashers for the crown wheel securing bolts.

The Carrier—To Replace

With the bevel pinion in place in the gear carrier and the pinion bearings correctly adjusted for pre-load as described in an earlier paragraph, the crown wheel and differential unit can be refitted to the carrier.

Replace the differential unit complete into the carrier and secure the end caps with the four nuts and spring washers. The differential unit bearings should be a tight fit into their respective housings in the carrier, since in manufacture the machining tolerances of the bearing housings are adjusted to give a .002-in. pre-load on the bearings.

The backlash between the mesh of the pinion and crown wheel teeth should be between .005 and .008 inch and the correct figure for each set of gears will be found etched on the back of the crown wheel. Measure the backlash when the gears are secured in the carrier by using a dial gauge to register against the crown wheel teeth while the pinion is firmly held. If the backlash is found to be too great or too small, adjustment may be effected by moving the shim or shims, positioned between the differential case and the inner race of the differential

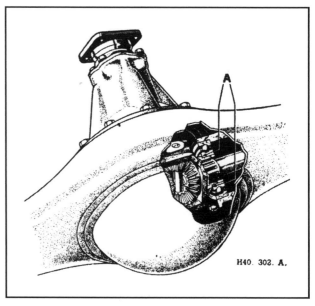

H40. 302. A.

Fig. 10. The backlash of .005–.008 in. between bevel pinion and crown wheel may be adjusted by shims A placed between differential case and crown wheel bearings.

bearings from one side of the crown wheel to the other as required.

The differential carrier can now be refitted into the axle case and secured by replacing the nuts. Use a new paper joint washer.

When all the bolts have been tightened the axle shafts (which are interchangeable from left to right) can be threaded through the hubs and secured on to the four-wheel studs as described earlier for the re-assembling of the hubs. When connecting up the propeller shaft use new lockwashers under the four nuts. Replace the axle drain plug and re-fill the unit with oil.

Replace the batteries and connect up the terminals.

Crown Wheel and Bevel Pinion Gears

During the manufacture of the crown wheel and bevel pinion gears, each crown wheel is carefully lapped-in with its corresponding pinion on a special running-in or gear-mating machine. After removal from this machine the two gears are carefully marked as a pair and they should never be separated and re-assembled individually in different axles. Should only a crown wheel or a pinion be damaged, a new pair of crown wheel and bevel pinion gears must be fitted if a satisfactory axle unit is to be built up.

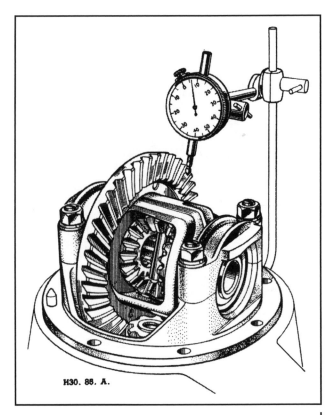

H30. 86. A.

Fig. 11. Making use of a dial indicator gauge to check the amount of backlash between crown wheel and pinion.

REAR SUSPENSION

SPRING DATA

	Type A	Type B	Type C
Free Length	$33\frac{1}{2}\pm\frac{1}{8}$ in.	$33\frac{1}{2}\pm\frac{1}{8}$ in.	$34\frac{3}{4}\pm\frac{1}{8}$ in.
Laden Length	$36\pm\frac{1}{8}$ in.	$36\pm\frac{1}{8}$ in.	$36\pm\frac{1}{8}$ in.
Free Camber	5 in.	$5\frac{1}{4}$ in.	$3\frac{3}{4}$ in.
Laden Camber	$\frac{1}{4}\pm\frac{1}{8}$ in. neg.	$\frac{1}{2}\pm\frac{1}{8}$ in. pos.	$\frac{1}{4}\pm\frac{1}{8}$ in. neg.
Deflection	$5\frac{1}{4}\pm\frac{1}{4}$ in. at 490 lbs.	$5\frac{1}{4}\pm\frac{1}{4}$ in. at 490 lbs.	$4\pm\frac{1}{4}$ in. at 490 lbs.
Number of leaves	7	7	8
Number of Zinc Interleaves	3	3	3

SPRINGS

Description

The rear semi-elliptic springs should be given regular attention, as the riding comfort of the car is largely dependent on their condition.

The springs have zinc interleaves between a number of their top leaves and they are also fitted with silentbloc bushes in the spring eyes. The shackle bearing to the frame is of the phosphor bronze type which requires periodic attention from the oil gun.

Periodic examination should be carried out to ensure that the spring leaf clips are tight and that none of the spring leaves are fractured.

The phosphor bronze bearings and the silentbloc bushes should also be checked for wear and the shackles must be examined for possible side play.

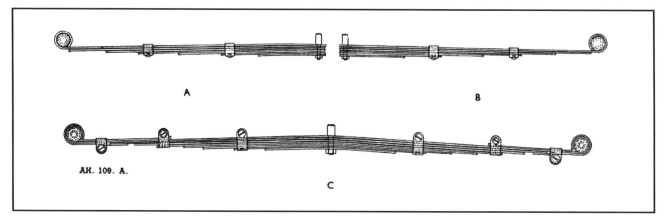

AH. 109. A.

Fig. 12. The three types of rear spring fitted to the Austin-Healey.
"C" is the latest type with wrap-over ends.

Removing a Shackle

Chock all the wheels except the one near the spring to be serviced. Jack the car until the rear wheel is clear of the ground and place suitable support blocks under the rear axle alongside the spring to axle securing clips. Gently lower the car on to the blocks and remove the rear wheel. If an under axle type of jack is available, it is better to place this under the spring securing clips instead of the blocks, as this permits the axle to be raised or lowered as required to relieve the spring of load.

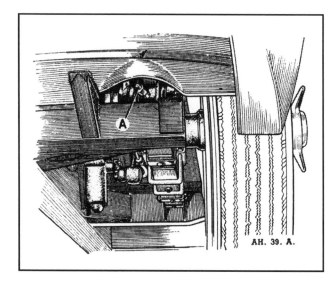

AH. 39. A.

Fig. 13. This illustration shows the position of the rear spring lubricator beneath the luggage compartment.

To remove the rear shackle complete, detach the nut and spring washer on the inside of the upper shackle and the locknut, spring washer and nut on the inside of the lower shackle. The shackle inside connecting link can now be removed and the top and bottom shackle pins together with the outside link pulled clear of their

respective bush bearings. The shackle assembly is then free.

To remove the shackle pin at the anchor or forward end of the spring remove the nut and spring washer on the inside of the pin and then drive the shackle pin clear. To give clearance for working and pin removal the exhaust pipe supporting bracket near this point should be removed.

Removing a Spring

First jack up the car on that side from which the spring is to be removed, then pack up to the chassis rear cross member with suitable supports, placing the supports as near to the spring rear anchorage as possible.

Place a screw jack under the centre of the spring to relieve the tension. Remove wheel as previously detailed, then using a box spanner release the four self-locking nuts from the spring clips ("U" bolts) which secure the spring to the axle tube.

If it is the left-hand spring that is being serviced, it will be found that on releasing the "U" bolts, the one

AH. 142. A.

Fig. 14. Spring rear shackle assembly in exploded form.

end of the anti-sway bar may be removed complete, see section "Anti-Sway Bar".

Release the front and rear shackle pins as previously described, but it may be found necessary during this operation to lower the screw jack slightly to gain access to the rear shackle pins.

Dismantling a Spring

Grip the spring in a vice, with the vice jaws against the top and bottom leaves, adjacent to the centre bolt. For spring type C of fig. 12 unscrew the clips.

For springs of pattern "A" and "B" open out the clips away from the spring leaves with suitable punch and hammer.

Finally, unscrew the nut from the centre pin and withdraw the pin. Carefully open the vice when the spring leaves, together with the zinc interleaving, will separate. These should now be thoroughly examined for signs of failure or cracks. Replace any defective leaves, thoroughly clean and regrease. For springs "A" and "B" ensure that the rivets of the clips are tight.

Replace the spring in a vice. Utilising a rod of similar diameter to the clamping bolt and having a tapered end, position the leaves so that the clamping bolt can be readily replaced without risk of damage to the thread.

Replace the clamping bolt and nut followed by the leaf clips of spring "C". New screwed pins may be necessary.

Renewal and Replacement of Shackle Bushes

Spring Eye Bushes: These are of the silentbloc type and must therefore be pushed clear of the spring eyes by applying pressure to the outer bush of the assembly. Again when replacing the silentbloc, pressure must only be applied to the outside bush. When the shackle pin is inserted the nut must be pulled up tight otherwise the silentbloc bush will not operate properly.

Shackle Bracket Bushes: In the spring rear mounting bracket there are two bronze bushes with sufficient space between to provide an oil channel for lubricant forced in through the nipple provided.

Should these bushes become worn and require re-placing, the shackle pins must be withdrawn, as already described, when the inner bush can be driven out of the mounting using a suitable punch and hammer. The nose of the punch should be inclined at an angle in order that it should engage with the end of the bush at the oil channel.

In order to withdraw the outer bush, a suitable extractor will have to be improvised. The following is a suggested method for bush removal, materials required being :—

(1). Bolt of 4 in.-length under head, screwed $\frac{3}{8}$ in. diameter its entire length.

(2). 3 nuts for bolt.

(3). Tube 1 in. long × $1\frac{1}{16}$ in. internal diameter.

(4). Tube 1 in. long × $\frac{13}{16}$ in. external diameter.

(5). Washer of $1\frac{7}{16}$ in. approximately external diameter.

(6). Washer of $\frac{13}{16}$ in. maximum external diameter.

Screw one nut up to the head of the bolt then pass large washer over bolt followed by large tube. Thread

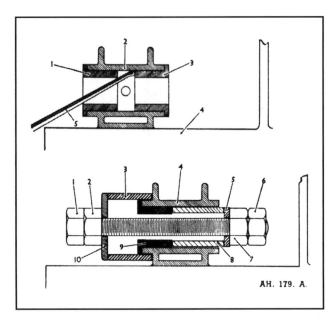

Fig. 15. Removing shackle bracket bushes.

Top. Drifting out the inner bush with:—1. Outer bush. 2. Bracket. 3. Inner bush. 4. Frame member. 5. Drift.

Bottom. Using the extractor described to remove the outer bush:— 1. Bolt. 2. Nut. 3. Large tube. 4. Bracket. 5. Small washer. 6. Locknut. 7. Nut. 8. Small tube. 9. Outer bush. 10. Large washer.

bolt through bush and bracket so that tube passes over bush flange to rest against bracket. Fit small tube into bracket from the inside of the chassis so that the bolt passes through its centre. Pass small washer over pro-truding bolt and screw up a second nut and secure with locknut.

By holding the bolt head with a spanner and turning the nut under its head down the bolt with a second span-ner the bush will be extracted from the bracket.

To fit replacement bushes, enter both bushes into the bracket then pass a bolt, with a large washer under its head, through the bushes and bracket from the outside of the bracket. Place a second large washer over the protruding bolt end on the inside of the bracket and screw a nut up on the thread until both bushes have been pulled firmly home. Remove the bolt.

Replacing a Spring

When the spring is fully assembled it should be fitted first at the anchor end and then at the shackle end. Remember that the shackle nuts must be pulled up tight if the silentbloc bush in the spring eye is to function correctly.

Jack up the spring to meet the axle, making sure that the aluminium packing piece is in place (tapered end pointing rearward) also the fibre pad. Replace the "U" bolts and tighten down the nuts on to the axle bracket with the serrated lockwashers beneath the nuts.

If it is the left-hand rear spring that is being refitted do not forget to fit the anti-sway bar and its bracket in place, also the exhaust pipe.

SHOCK ABSORBERS

Description

The shock absorbers are Armstrong double-acting hydraulic, resistance being offered to the compression and to the recoil of the road springs.

Shock absorber maintenance is confined to the periodical examination of the anchorage to the chassis, the two fixing bolts being tightened as required, and topping-up with fluid.

No adjustment is required or provided for, and any attempt to dismantle the piston assembly will seriously affect the performance of the shock absorber.

Topping-up with Fluid

Before removing the filler plug carefully wipe clean the exterior of the shock absorber body. This is most important, as it is vital that dirt or foreign matter on no account enters the interior of the unit.

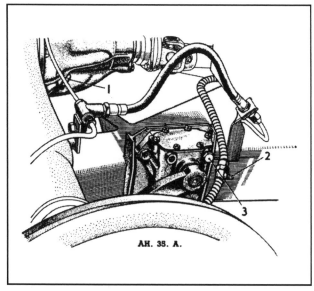

Fig. 16. *Chassis rear end lubrication.*
1. *Axle filler plug.* 2. *Handbrake cable nipple.* 3. *Rear shock absorber filler plug.*

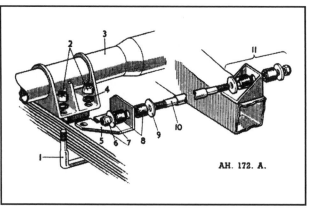

AH. 172. A.

Fig. 17. *The anti-sway bar exploded.*

1. *Spring "U" bolt.*	6. *Self-locking nut.*
2. *"U" bolt locking nut.*	7. *Washer.*
3. *Axle tube.*	8. *Rubber bushes.*
4. *Serrated washer.*	9. *Washer.*
5. *Anti-sway bar anchor plate.*	10. *Anti-sway bar.*

11. *R.H. anchorage assembly.*

Ensure that only Armstrong recommended shock absorber fluid is used for topping-up.

While adding fluid the lever arm must be worked throughout its full stroke to expel any air that might be present in the working chamber.

Fluid should be added to the level of the bottom of the filler plug hole.

Testing

When there is any doubt that the rear suspension of the car is not adequately damped, the conditions of the road springs and tyre pressures should be borne in mind.

If the shock absorbers do not appear to function satisfactorily, an indication of their resistance can be obtained by carrying out the following check :

It is advisable to remove the shock absorbers from the chassis, but testing can be done in position providing the axle link is disconnected at the shock absorber arm. If removed, place individually in a vice, taking care to grip by the fixing lugs, to avoid distortion of the cylinder body.

Move the lever arm up and down through its complete stroke. A moderate resistance throughout the full stroke should be felt.

If resistance is erratic, and free movement of the lever arm is noted, it may indicate lack of fluid. If the addition of fluid as described, gives no improvement a new or reconditioned unit must be fitted.

Too much resistance, when it is not possible to move the lever arm slowly by hand, may indicate a broken internal part or a seized piston in which case the unit will have to be replaced.

Removing Shock Absorber

Remove the nut and spring washer that secures the shock absorber lever to the link arm between lever and axle. Withdraw the two fixing setpins from the shock absorber body and chassis bracket, then remove the shock absorber, threading the lever over the link arm bolt.

Replacing Shock Absorber

Shock absorbers may be replaced by simply reversing the procedure outlined for their removal. However, when handling shock absorbers that have been removed from the chassis for any purpose, it is important to keep the assemblies upright as much as possible otherwise air may enter the working chamber and so cause erratic resistance.

Connecting Link Bushes

The rubber bushes integral with both ends of the connecting link which joins the shock absorbers to the rear axle cannot be renewed. When these bushes are worn the arm must be renewed complete.

Anti-Sway Bar

The anti-sway bar is a torsion bar anchored by a bracket clamped by the left hand spring "U" bolts and at the right-hand chassis member by a bracket welded to the frame. The assembly of the rubber bushes and anti-sway bar is clearly shown in fig. 17.

SERVICE DIAGNOSIS GUIDE

Symptom	No.	Possible Cause
(a) Noisy Axle **(i) On drive**	1	Insufficient crown wheel/pinion backlash
(ii) On coast	2	Excessive crown wheel/pinion backlash
(iii) On drive and coast	3	Insufficient lubricant
	4	Chipped, broken or scored teeth
	5	Damaged or worn bearings
(b) Broken Crown Wheel and Pinion Teeth	1	Excessive loads
	2	Clutch snatch
	3	Incorrect backlash adjustment
	4	Incorrect differential bearing adjustment (preload)
(c) Broken or Scored Differential Gear Teeth		In (a), check 3
		In (b), check 2
	1	Misaligned or bent differential shafts
	2	Badly worn differential wheel and pinion thrust washers
(d) Differential Shaft Breakage		In (b), check 1 and 2
(e) Overheating		In (a), check 1 and 3
	1	Bearings adjusted too tightly
(f) Oil Leaks	1	Damaged gear carrier flange
	2	Damaged or broken joint washers
	3	Worn or damaged oil seals

SERVICE JOURNAL REFERENCE

NUMBER	DATE	SUBJECT	CHANGES

BRAKES

GENERAL DATA

Make	Girling
Type	Hydraulic, two leading shoes
Pedal free movement...	$\frac{1}{8}$-in. (3.175 mm.)
Handbrake	Mechanical, rear wheels only
Drum diameter	11-in. (28 cm.)
Total frictional area	142 sq. in. (916 sq. cm.)
Shoe lining width	1$\frac{3}{4}$-in. (44.45 mm.)
Shoe lining length :	
Front	10$\frac{3}{4}$-in. (27.3 cm.)
Rear leading shoe	10$\frac{3}{4}$-in. (27.3 cm.)
Rear trailing shoe	9$\frac{5}{8}$-in. (24.45 cm.)
Shoe lining thickness...	$\frac{3}{16}$-in. (4.76 mm.)

Description

The brakes on all four wheels are hydraulically operated by a foot pedal coupled to a master cylinder in which the hydraulic pressure of the brake operating fluid is originated. A supply tank provides a reservoir by which the fluid is replenished, and a pipe line consisting of tube, flexible hose and unions, interconnect the supply tank, master cylinder and the wheel cylinders.

The pressure generated in the master cylinder by application of the foot pedal is transmitted with equal and undiminished force to all wheel cylinders simultaneously. This moves the pistons outwards, which in turn expand the brake shoes, thus producing automatic equalization and efficiency in direct proportion to the effort applied at the pedal.

When the pedal is released the brake shoe springs return the wheel cylinder pistons, and therefore, the fluid back into the system.

An independent mechanical linkage actuated by a handbrake of pull-up type, operates the rear wheels by mechanical expanders attached to the rear wheel cylinder bodies.

The front brakes are of the two leading shoe type, with sliding shoes ensuring automatic centralization of the brake shoes in operation.

The rear brakes are of the non servo type, with sliding shoes, and incorporate the handbrake mechanism.

Front Brakes

The front brake shoes are operated by two wheel cylinders situated diametrically opposite each other on the inside of the backplate and interconnected by a bridge pipe on the outside.

Each wheel cylinder consists of a light alloy body containing a spring, seal support, seal, steel piston, and a rubber dust cover. The inclined abutment (28 deg.) at the base of the body has a steel lined slot to avoid wear.

Each brake shoe is located in the slot in the base of one wheel cylinder and expanded by the piston of the other, with the leading edges of both shoes making initial contact with the drum. The shoes are allowed to slide and centralize during the actual braking operation

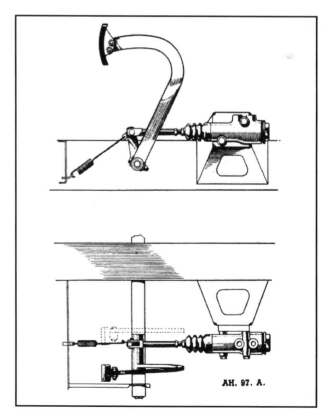

Fig. 1. *Showing the brake pedal and master cylinder layout for right-hand drive models.*

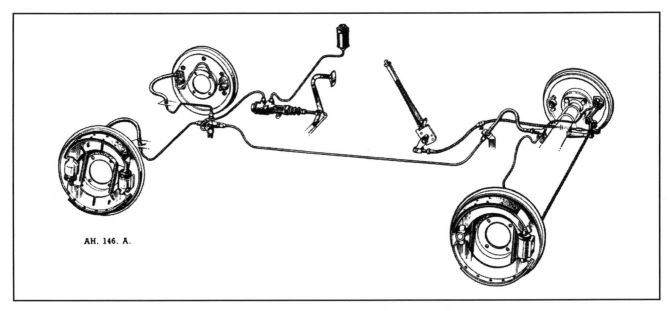

AH. 146. A.

*Fig. 2. A general arrangement of the braking system showing
both the full hydraulic and the parking brake layouts.*

which distributes the braking force equally over the lining area, ensuring high efficiency and even lining wear.

Adjustment for lining wear is by means of two knurled snail cam adjusters, each operating against a peg at the actuating end of each shoe. Both adjusters turn clockwise to expand the shoes.

The shoes are held in position by two return springs which pass from shoe to backplate. One adjustable steady post is located under each shoe to allow the shoes to be set square on the backplate and in the drums, felt bushes are fitted to the posts to act as lubricant retainers.

A bleed screw incorporated in one wheel cylinder presses a steel ball firmly on a valve aperture in the cylinder and a rubber cap is fitted to exclude dirt.

The shoes are held in position by two shoe return springs fitted from shoe to shoe between shoe and backplate with the shorter spring nearer the adjuster.

The adjuster has a steel housing which is spigoted and bolted firmly to the inside of the backplate.

The housing carries two opposed steel links, the outer ends of which are slotted to take the shoes, whilst the inclined inner faces bearing on the corresponding sloping faces of the hardened steel wedge (the axis of which is at right angles to the links).

The wedge has a finely threaded spindle with a square end which projects on the outside of the backplate. By rotating the wedge in a clockwise direction, the links are forced apart, and the fulcrum of the brake shoes expanded.

Rear Brakes

The rear brake shoes are not fixed but allowed to slide and centralize with same effect as in the front brakes. They are hydraulically operated by a double acting wheel cylinder, and lining wear is adjusted by a wedge type mechanical adjuster common to both shoes.

The cylinder and adjuster are situated opposite one another on the inside of the backplate with the ends of the shoes locating in slots in the wheel cylinder pistons and adjuster links respectively.

Each shoe rests on an adjustable steady post having a felt bush.

The wheel cylinder consists of a die-cast aluminium housing containing a centrally located spring at either end of which is situated a seal support, a seal, and a piston with dust cover.

A bleed screw is also incorporated in the cylinder housing with a rubber dust cap over the nipple end.

The nuts securing the wheel cylinder to the backplate should be half to one turn slack to allow for centralization of the handbrake mechanism.

The rear brakes are mechanically operated by a handbrake mechanism housed on the side of the wheel cylinder casting, and consists of two flat inclined faced hardened steel tappets (which expand the shoes), two hardened steel rollers, and a flat hardened steel wedge which acts as a draw link. The retaining cover is secured to the housing by four setscrews and has two tabs to prevent the tappets sliding out of the housing when the brake shoes are removed.

Handbrake

The hand or parking brake operates on the rear wheels only and is applied by a control situated between the two seats. The cable from the control is attached to the compensator mounted on the rear axle, transverse rods, which are non-adjustable, connect the compensator with the draw link of the expander housing.

The handbrake linkage is set when leaving the works and should not require any alteration. Should the linkage have been disturbed or the normal running adjustment at the brake shoes be insufficient, then the following procedure should be adopted.

Lock up the shoes in the drums by means of the adjusters, apply the handbrake one notch and adjust the cable from control to compensator, to take up any slackness. Adjust the brakes as described.

MASTER CYLINDER

This is of the compression type where the operation of the brake pedal pushes the plunger into the compression chamber. The cylinder is fixed to the chassis frame by two bolts.

The assembly consists of a cast iron housing with highly finished bore into which is assembled the polished steel plunger, plunger return spring, recuperating seal which has two internal diameters, a steel shim to protect the seal, outer or end seal and seal retainer. The screwed end cap is machined to support the recuperating seal.

Operation of the plunger is provided by means of a push rod with a spherical end, seating in the concave end of the seal retainer. To protect the operating end of the cylinder from dirt and dust, etc., a rubber boot or bellows is provided, which in all cases should be packed with Girling Rubber Grease No. 3 (RED).

Dismantling

Before removing the master cylinder for dismantling it is advisable to drain off most of the brake fluid by disconnecting one of the flexible brake hoses from one wheel cylinder, lowering the open end into a clean receptacle and pumping the foot pedal until no further fluid enters the container. Re-connect the hose.

Disconnect the two pipe unions on the top of the cylinder and the operating rod from its connection to the foot pedal. Withdraw the two securing bolts and remove the cylinder.

First unscrew the end cap and remove complete with gasket, followed by the removal of the plunger return spring. Remove the rubber boot, and by using a pair of taper nosed pliers withdraw the circlip and the operating rod complete is then detached from the cylinder. The plunger, complete with seal retainer and end seal is pushed from the pressure end of the cylinder. Remove the recuperating seal followed by a steel shim.

Carefully examine the various parts and renew any that appear worn or damaged. It is important to renew any of the seals which are perished, worn, or distorted.

Assembling

After thoroughly cleaning all parts with clean brake fluid, place the steel shim against the shoulder which is formed inside the body and replace the recuperating seal with the back towards the shim, making sure that it is correctly seated. Assemble the end seal, with the wider end first on to the shoulder machined on the plunger, and insert the seal retainer into the hole provided. Insert the plunger assembly, with open end first, from the operating end of the cylinder.

Re-assemble operating rod washer and circlip, pack the boot with Girling Rubber Grease No. 3 (Red), and stretch over the end of the cylinder. Replace the plunger

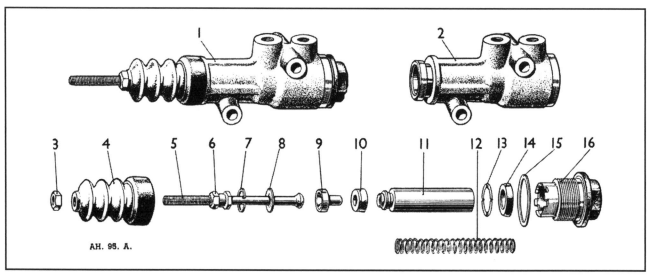

Fig. 3. Components of the Master Cylinder.

1. *Master cylinder complete.*
2. *Master cylinder body.*
3. *Locknut.*
4. *Rubber boot.*
5. *Push rod.*
6. *Collar.*
7. *Circlip.*
8. *Retaining washer.*
9. *Seal washer.*
10. *Seal.*
11. *Plunger.*
12. *Return spring.*
13. *Shim washer.*
14. *Recuperating seat.*
15. *Gasket.*
16. *End cover.*

return spring and end cap complete with gasket, and tighten firmly.

Refit the master cylinder to the chassis in reverse order to removal.

Note :—It is advisable that the rubber seals and the polished steel plunger should be smeared with clean brake fluid before assembly.

On no account should petrol, paraffin, or trichoethylene be used for cleaning any of the hydraulic parts. Only pure methylated spirits or clean brake fluid should be employed for this purpose. When in operation, the plunger must **never** be prevented from returning to its full "off" position.

With the pedal pad held in its own full "off" position. adjust the master cylinder operating rod to allow $\frac{1}{8}$-in. free movement at the pedal pad.

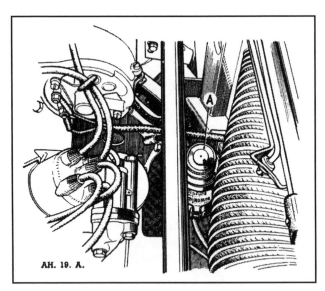

AH. 19. A.

Fig. 4. Showing the position of the brake fluid supply tank for right-hand drive models.

Supply Tank

The type fitted consists of a cylindrical metal container with a screwed metal cap pierced with a breather hole (which must always be kept clear) and containing a cap washer. The tank is fixed to the scuttle panel for left-hand drive cars and to the right-hand engine mounting bracket for right-hand drive models, by a metal clip, nut and bolt, and connected to the master cylinder by unions and a pipe.

FITTING REPLACEMENT BRAKE SHOES

Always fit Girling "Factory Lined" replacement shoes. These have the correct type of lining, and are accurately ground to size which ensures a fast and easy bed in the drums.

When fitting replacement shoes, always fit a new set of shoe return springs.

The following instructions if carefully carried out in sequence should present no difficulty to either owner or mechanic.

Front Brakes

1. Jack up the car and remove road wheels and brake drums.
2. Lift one shoe out of the abutment slot of one wheel cylinder, then release from the piston slot of the other. (It will be found quite simple to remove the shoe return springs.)
   To prevent the wheel cylinder pistons from expanding it is advisable to place a rubber band round each cylinder. Repeat with the second shoe.
3. Clean down the backplate, check wheel cylinders for leaks and freedom of motion.
4. Check adjusters for easy working and turn back (anti-clockwise) to full "off" positions. Lubricate where necessary with **Girling (White) Brake Grease.**
5. Smear the tips of the steady posts, the operating and abutment ends of the new shoes with **Girling (White) Brake Grease.**
   Girling (white) brake grease must not be allowed to contact hydraulic cylinders, pistons or rubber parts. Keep all grease off the linings on new replacement shoes and do not handle lining more than necessary.
6. Fit new shoe return springs to the new shoes. Place one end of the spring through the hole in the shoe web and the other end through the hole in the back-

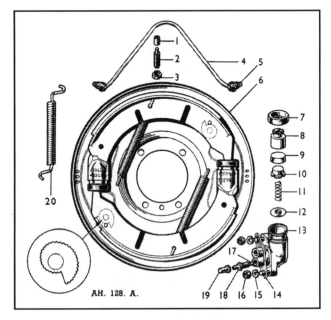

AH. 128. A.

Fig. 5. Front brake backplate.

1. Felt bush.
2. Steady post.
3. Locknut.
4. Cylinder connecting pipe.
5. Pipe union.
6. Backplate.
7. Rubber dust cover.
8. Piston.
9. Seal.
10. Seal support.
11. Spring.
12. Spring locating plate.
13. Cylinder housing.
14. Joint washer.
15. Washer.
16. Nut.
17. Bleed valve ball.
18. Bleed screw.
19. Bleed screw cover.
20. Shoe return spring.

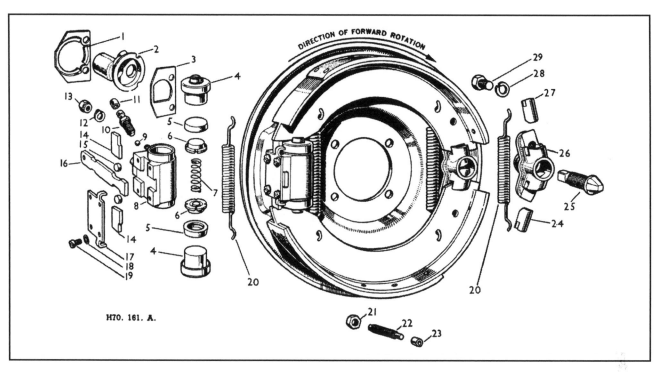

H70. 161. A.

Fig. 6. Rear brake backplate.

1. Dust cover retainer plate.
2. Dust cover.
3. Distance piece.
4. Piston.
5. Seal.
6. Seal support.
7. Spring.
8. Cylinder body.
9. Bleed valve ball.
10. Bleed screw.
11. Bleed screw cover.
12. Washer.
13. Nut.
14. Tappet.
15. Roller.
16. Roller wedge.
17. Cover plate.
18. Washer.
19. Setscrew.
20. Shoe return spring.
21. Locknut.
22. Steady post.
23. Felt bush.
24. Link.
25. Adjuster wedge.
26. Adjuster body.
27. Link.
28. Spring washer.
29. Set bolt.

plate near the abutment end of the same shoe. Each shoe can be replaced independently. Remove rubber bands from cylinder.

7. Make sure the drums are clean and free from grease, etc., then refit.
8. Adjust the brakes as described below.
9. Refit the road wheels and lower the car to the ground.

Rear Brakes

Proceed in stages as described for "Front Brakes", Paras. 1-9, substituting the details in the following paragraphs for those bearing the same number.

2. Lift one of the shoes out of the slots in the adjuster link and wheel cylinder piston. Both shoes can then be removed complete with springs. Place a rubber band round the wheel cylinder to keep the piston in place.

6. Fit the two new shoe return springs to the new shoes (with the shorter spring at the adjuster end) from shoe to shoe and between the shoe web and the backplate. Locate one shoe in the adjuster link and wheel cylinder piston slots, then prise over the opposite shoe into its relative position. Remove the rubber band.

Note :—The first shoe has the lining positioned towards the heel of the shoe and on the second shoe towards the toe or operating end in both LH and RH brake assemblies.

Several hard applications of the pedal should be made to ensure all the parts are working satisfactorily and the shoes bedding to the drums, then the brakes should be adjusted as described.

RUNNING MAINTENANCE

The Brakes are adjusted for lining wear **Only** at the brakes themselves and on no account should any alteration be made to the handbrake cable for this purpose.

Front Brakes

A separate snail cam adjuster is provided for each shoe. Jack up the car until the front wheel to be adjusted is clear of the ground, then fully release both hexagon head adjuster bolts on the outside of the backplate, by turning anti-clockwise with an open ended spanner.

Turn one of the adjuster bolts clockwise until the brake shoe concerned touches the brake drum, then release the adjuster until the shoe is just free of the drum. Repeat the process for the second adjuster and shoe.

Spin the wheel to ensure that the brake shoes are quite free of the drum. Repeat the whole procedure for the second front wheel.

Rear Brakes

One common adjuster is provided for both shoes and the adjustment of both rear wheels is identical.

Release the handbrake and jack up the car. Turn the square end of the adjuster on the outside of each rear brake backplate in a clockwise direction until resistance is felt. Slacken two clicks when the drum should rotate freely.

Immediately after fitting replacement shoes, it is advisable to slacken one further click to allow for possible lining expansion, reverting to normal adjustment afterwards.

Steady Posts

Should the steady posts require resetting, first slacken the steady post lock nuts and turn the posts anti-clockwise with a screwdriver. Lock the shoes up in the drums by the adjusters until tight, then turn the steady posts clockwise until they bear against the shoes. Re-tighten lock nuts and adjust the brakes as described.

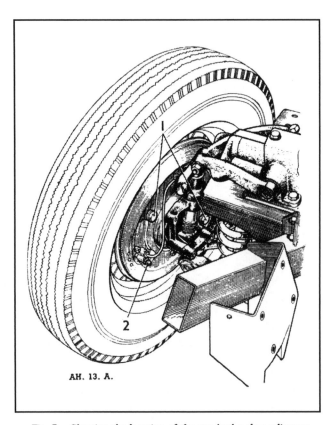

AH. 13. A.

Fig. 7. Showing the location of the two brake shoe adjusters and the brake bleed nipple on a front brake backplate.

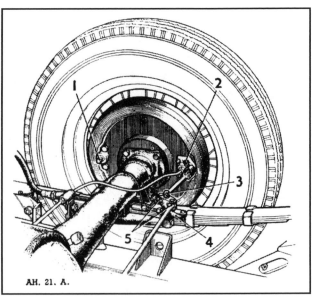

AH. 21. A.

Fig. 8. A rear brake backplate.

1. *Brake shoe adjuster.*   2. *Bleed nipple.*
3. *Balance lever lubricator.*   4. *Brake rod oiling point.*
       5. *Transverse rod oiling points.*

GENERAL MAINTENANCE

Replenishment of Hydraulic Fluid

Inspect the supply tank at regular intervals and maintain at about three quarters full by the addition of **Girling Crimson Brake Fluid.**

Great care should be exercised when adding brake fluid to prevent dirt or foreign matter entering the system.

Important: Serious consequences may result from the use of incorrect fluids and on no account should any other than Girling Crimson Brake Fluid be used. This fluid has been specially prepared and is unaffected by high temperature or freezing.

Bleeding the Hydraulic System

Bleeding is necessary any time a portion of the hydraulic system has been disconnected, or if the level of the brake fluid has been allowed to fall so low that air has entered the master cylinder.

With all the hydraulic connections secure and the supply tank topped up with fluid, remove the rubber cap from the L.H. rear bleed nipple and fit the bleed tube over the bleed nipple, immersing the free end of the tube in a clean jar containing a little brake fluid.

Unscrew the bleed nipple about three-quarters of a turn and then operate the brake pedal with a slow full stroke until the fluid entering the jar is completely free of air bubbles. Then during a down stroke of the brake pedal, tighten the bleed screw sufficiently to seat the ball, remove bleed tube and replace the bleed nipple dust cap. **Under no circumstances must excessive force be used when tightening the bleed screw.**

This process must now be repeated for each bleed screw at each of the three remaining backplates, finishing at the wheel nearest the master cylinder. Always keep a careful check on the supply tank during bleeding, since it is most important that a full level is maintained. Should air reach the master cylinder from the supply tank, the whole of the bleeding operation must be repeated.

After bleeding, top up the supply tank to its correct level of approximately three-quarters full.

Never use fluid that has just been bled from a brake system for topping up the supply tank, as this brake fluid may be to some extent aerated. Such fluid must be allowed to stand for at **least** twenty-four hours before it is used again. This will allow the air bubbles in the fluid time to disperse.

Great cleanliness is essential when dealing with any part of the hydraulic system, and especially so where the brake fluid is concerned. Dirty fluid must never be added to the system.

Note:—It is advisable to turn all the brake shoe adjusters to their full "off" position before bleeding. After bleeding adjust brakes as described.

GENERAL ADVICE

Always exercise extreme cleanliness when dealing with any part of the hydraulic system.

Always take care not to scratch the highly finished surfaces of cylinder bores or pistons.

Always use clean Girling Brake Fluid or alcohol for cleaning internal parts of the hydraulic system. On no account should petrol or paraffin be allowed to contact these parts.

Always examine all seals carefully when overhauling hydraulic cylinders, and replace with **Genuine Girling Spares** any which show the least sign of wear or damage.

Always use **Girling Crimson Brake Fluid.**

Always use **Girling Rubber Grease No. 3 (Red)** for packing rubber boots and dust covers and lubricating parts likely to contact any rubber components. **Never** use Girling White Brake Grease for these purposes.

Important : If it is suspected that incorrect fluids have been used **All Seals** in the master cylinder and wheel cylinders must be changed after the components and pipe-lines have been thoroughly flushed and cleaned out with alcohol or clean **Girling Crimson Brake Fluid. Never use Petrol or Paraffin for this purpose.**

If incorrect fluid has been in the system for any length of time, it is advisable to replace the high pressure hoses.

SERVICE DIAGNOSIS GUIDE

Symptom	No.	Possible Fault
(a) Spongy Pedal **(loss of fluid pressure)**	1 2 3 4 5 6	Leak in system Master cylinder main cup worn Master cylinder secondary cup worn Wheel cylinder leaking Air in system Lining not "down" on shoe
(b) Excessive Pedal Depression	 1 2 3	In (a) check 1 and 5 Excessive lining wear Extremely low brake fluid level Too much pedal free movement
(c) Brakes Grab or Pull to Side	1 2 3 4 5 6 7 8 9 10	Brake backplate loose on axle Scored, cracked or distorted drum High spots on drum Incorrect shoe adjustment Oily or wet linings Rear axle or front suspension anchorage loose Worn or loose rear spring anchorage Worn steering connections Different grades or types of lining fitted Uneven tyre pressures
(d) Dragging Brakes	 1 2 3 4 5 6 7	In (c) check 3 Wheel cylinder piston seized Weak or broken brake shoe return springs Master cylinder by-pass port restricted Too little pedal free movement Handbrake mechanism seized Supply tank overfilled Filler cap air vent choked
(e) Springy Pedal	1 2 3	Linings not "bedded-in" Brake drums weak or cracked Master cylinder fixing loose
(f) Brakes Inefficient	 1	In (c) check 4 In (d) check 7 Incorrect type of linings fitted

WHEELS AND TYRES

A MOST important factor in the road-worthiness of the car is systematic and correct tyre maintenance. Tyres must be able to sustain the weight of a loaded car and be able to withstand the vagaries of road conditions. Pressures should be checked at least once each week.

Both front and rear tyres nearer to the curb should be inflated with a pressure two or three pounds per square inch above the pressure in the tyres nearest the crown of the road. The benefit of this differential pressure will be felt in easier handling and less tyre wear, particularly where roads are winding, and are heavily, or even moderately, cambered.

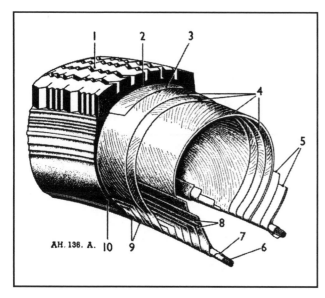

Fig. 1. Diagrammatic illustration of tyre construction.

1. *Tread.*
2. *Soft Cushion Rubber.*
3. *Breaker.*
4. *Casing Plies.*
5. *Fillers.*
6. *Bead Wires.*
7. *Bead Wrapping.*
8. *Fillers.*
9. *Chafers.*
10. *Wall Rubber.*

GENERAL DESCRIPTION

The Austin-Healey '100' car is fitted with 5.90–15 Dunlop Road Speed tyres upon wire spoked wheels with knock-on type hub caps.

These tyres have been specially designed for high performance cars, having tread and casing compounds to resist heat generation.

Special attention is directed to the inflation pressure recommendations. One of the principal functions of the tyres fitted to a car is to eliminate high frequency vibrations. They do this by virtue of the fact that the unsprung mass of each tyre—the part of the tyre in contact with the ground—is very small.

Tyres must be flexible and responsive. They must also be strong and tough to contain the air pressure, resist damage, give long mileage, transmit driving and braking forces, and at the same time provide road grip, stability and good steering properties.

Strength and resistance to wear are achieved by building the casing from several plies of cord fabric, secured at the rim position by wire bead cores and adding a tough rubber tread, fig. 1.

Part of the work done in deflecting the tyres on a moving car is converted into heat within the tyres. Rubber and fabric are poor conductors and internal heat is not easily dissipated. Excessive temperature weakens the tyre structure and reduces the resistance of the tread to abrasion by the road surface.

Heat generation, comfort, stability, power consumption, rate of tread wear, steering properties and other factors affecting the performance of the tyres and car are associated with the degree of tyre deflection. All tyres are designed to run at predetermined deflections, depending upon their size and purpose.

INFLATION PRESSURES

Operating Conditions	Pressure in lb/sq. in.	
	Front	Rear
Normal motoring in Great Britain and similar road and traffic conditions elsewhere (when sustained high speeds are not possible) ...	20	23
Continental type touring with lengthy periods at sustained speeds in excess of 80-90 m.p.h. (128-144 k.p.h.) ...	26	29
Motoring which is predominantly and regularly of the high speed touring type	28	31

Inflation pressures for Road Speed tyres vary with conditions of use. The foregoing recommendations apply to cars used under ordinary road conditions, either in U.K. or overseas. When cars are to be used for racing or special high speed testing where sustained speed of more than 110 m.p.h. is anticipated, it is desirable to consult either the Service Department of The Austin Motor Co. Ltd., or the nearest depot of the Dunlop Rubber Company, as to the need for tyres of full racing construction, and the correct pressures to be used.

Pressures should be checked when the tyres are cold, such as after standing overnight, and not when they have attained normal running temperatures.

It should be remembered that tyres lose pressure, even when in sound condition, due to a chemical diffusion of the compressed air through the tube walls. The rate of loss in a sound car tyre is usually between 1 lb. and 3 lbs. per week. For this reason and with the additional purpose of detecting slow punctures, pressure should be checked with a tyre gauge applied to the valve not less often than once per week.

Any unusual pressure loss should be investigated immediately. Having first made sure that the valve is not leaking, remove the tube and water test.

Do not over inflate a tyre and do not reduce pressures which have increased owing to increased temperature.

Valve cores are inexpensive and it is a wise precaution to renew them periodically. Valve caps should always be fitted and renewed when the rubber seatings have become damaged after constant use.

TYRE EXAMINATION

Tyres on cars submitted for servicing should be examined for :—
(1) Inflation pressures.
(2) Degree and regularity of tread wear.
(3) Misalignment.
(4) Cuts and penetrations.
(5) Small objects embedded in the treads, such as flints and nails.
(6) Impact bruises.
(7) Kerb damage on walls and shoulders.
(8) Oil and grease.
(9) Contact with the car.

Oil and grease should be removed by using petrol sparingly. Paraffin is not sufficiently volatile and not recommended. If oil or grease on the tyres results from over lubrication or defective oil seals, suitable corrections should be made.

REPAIR OF INJURIES

Minor injuries confined to the tread rubber, such as from small pieces of glass or road dressing material,

require no attention other than the removal of the objects. Cold filling compound or "stopping" is unnecessary in such cases.

More severe tread cuts and wall damage, particularly if they penetrate to the outer ply of the fabric casing, require vulcanised repairs. The Dunlop Spot Vulcanising Unit is available for this purpose and it is also suitable for all types of tube repairs.

Injuries which extend into or through the casing, except clean nail holes, seriously weaken the tyre. Satisfactory repair necessitates new fabric being built in and vulcanised. This requires expensive plant and should be undertaken by a tyre repair specialist.

Loose gaiters and "stick-in" fabric repair patches are not satisfactory substitutes for vulcanised repairs and should be used only as a temporary "get-you-home" measure if the tyre has any appreciable tread remaining. They can often be used successfully in tyres which are nearly worn out and which are not worth the cost of vulcanised repairs.

Clean nail holes do not necessitate cover repairs. If a nail has penetrated the cover the hole should be sealed by a tube patch attached to the inside of the casing. This will protect the tube from possible chafing at that point.

AH. 163. A.

Fig. 2. Excessive tyre distortion from persistent underinflation causes rapid wear on the shoulders and leaves the centre standing proud. If the effects of under inflation are aggravated by other factors, such as camber and excessive braking, the irregular and rapid wear is more pronounced.

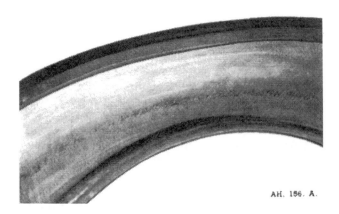

AH. 186. A.

Fig. 3. This casing is breaking up due to over-flexing and heat generation.

If nail holes are not clean, and particularly if frayed or fractured cords are visible inside the tyre, expert advice should be sought.

FACTORS AFFECTING TYRE LIFE AND PERFORMANCE

Inflation Pressures

All other conditions being favourable there is an average loss of 13% tread mileage for every 10% reduction in inflation pressure below the recommended figure.

A tyre is designed so that there is a minimum pattern shuffle on the road surface and a suitable distribution of load over the tyre's contact area when deflection is correct.

Moderate underinflation causes an increased rate of tread wear although the tyre's appearance may remain normal. Severe and persistent underinflation produces unmistakable evidence on the tread, see fig. 2. It also causes structural failure due to excessive friction and temperature within the casing, figs. 3 and 4.

Pressures which are higher than those recommended for the car reduce comfort. They may also reduce tread life due to a concentration of the load and wear on a smaller area of tread aggravated, by increased wheel bounce on uneven road surfaces. Excessive pressures overstrain the casing cords, in addition to causing rapid wear, and the tyres are more susceptible to impact fractures and cuts.

Effect of Temperature

Air expands with heating and tyre pressures increase as the tyres warm up. Pressures increase more in hot weather than in cold weather and as a result of high speed. These factors are taken into account when designing the tyre and in preparing Load Pressure schedules.

Pressures in warm tyres should not be reduced to standard pressure for cold tyres. "Bleeding" the tyres increases their deflections and causes their temperatures to climb still higher. The tyres will also be underinflated when they have cooled.

Speed

High speed is expensive and the rate of wear tread may be twice as fast at 50 m.p.h. as at 30 m.p.h.

High speed involves :—
(1) Increased tyre temperatures due to more deflections per minute and a faster rate of deflection and recovery. The resistance of the tread to abrasion decreases with increase of temperature.
(2) Fierce acceleration and braking.
(3) More tyre distortion and slip when negotiating bends and corners.
(4) More "thrash" and "scuffing" from road surface irregularities.

Braking

"Driving on the brakes" increases the rate of tyre wear, apart from being generally undesirable. It is not necessary for wheels to be locked for an abnormal amount of tread rubber to be worn away.

Other braking factors not directly connected with the method of driving can affect tyre wear, for instance correct balance and lining clearances, and freedom from binding, are very important. Braking may vary between one wheel position and another due to oil or foreign matter on the shoes even when the brake mechanism is free and correctly balanced.

Brakes should be relined and drums reconditioned in complete sets. Tyre wear may be affected if shoes are relined with non-standard material having suitable characteristics or dimensions, especially if the linings differ between one wheel position and another in such a way

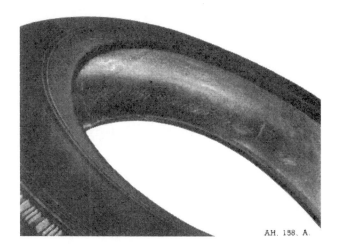

AH. 188. A.

Fig. 4. Running deflated has destroyed this cover.

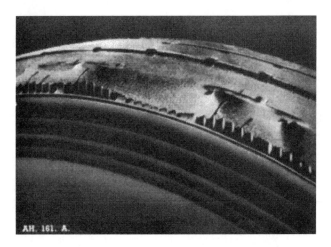

Fig. 5. Local excessive wear due to brake drum eccentricity.

as to upset the brake balance. Front tyres, and particularly near front tyres, are very sensitive to any condition which adds to the severity of front braking in relation to the rear.

"Picking-up" of shoe lining leading edges can cause grab and reduce tyre life. Local "pulling-up" or flats on the tread pattern can often be traced to brake drum eccentricity, fig. 5. The braking varies during each wheel revolution as the minor and major axis of the eccentric drum pass alternately over the shoes. Drums should be free from excessive scoring and be true when mounted on their hubs with the road wheels attached.

Climatic Conditions

The rate of tread wear during a reasonably dry and warm summer can be twice as great as during an average winter.

Water is a rubber lubricant and tread abrasion is much less on wet roads than on dry roads. In addition resistance of the tread to abrasion decreases with increase in temperature.

When a tyre is new its thickness and pattern depth are at their greatest. It follows that heat generation and pattern distortion due to flexing, cornering, driving and braking are greater than when the tyre is part worn.

Higher tread mileages will usually be obtained if new tyres are fitted in the autumn or winter rather than in the spring or summer. This practice also tends to reduce the risk of road delays because tyres are more easily cut and penetrated when they are wet than when they are dry. It is, therefore, advantageous to have maximum tread thickness during wet seasons of the year.

Road Surface

Present day roads generally have better non-skid surfaces than formerly. This factor, combined with improved car performance, has tended to cause faster tyre wear, although developments in tread compounds and patterns have done much to offset the full effects.

Road surfaces vary widely between one part of the country and another, often due to surfacing with local material. In some areas the surface dressing is coarser than others; the material may be comparatively harmless rounded gravel, or more abrasive crushed granite, or knife-edged flint. Examples of surfaces producing very slow tyre wear are smooth stone setts and wood blocks, but their non-skid properties are poor.

Bends and corners are severe on tyres because a car can be steered only by misaligning its wheels relative to the direction of the car. This condition applies to the rear tyres as well as the front tyres. The resulting tyre slip and distortion increase the rate of wear according to speed, load, road camber and other factors, fig. 6.

The effect of hills, causing increased driving and braking torques with which the tyres must cope, needs no elaboration.

Impact Fractures

In order to provide adequate strength, resistance to wear, stability, road grip and other necessary qualities, a tyre has a certain thickness and stiffness. Excessive and sudden local distortion, such as may result from striking a kerb, a large stone or brick, an upstanding manhole cover, or a deep pothole may fracture the casing cords; figs. 7 and 8.

Impact fractures often puzzle the car owner because the tyre and road spring may have absorbed the impact without his being aware of anything unusual. Only one or two casing cords may be fractured by the blow and the weakened tyre fails some time later. Generally there is no clear evidence on the outside of the tyre unless the object has been sufficiently sharp to cut it.

This damage is not associated solely with speed and care should be exercised at all times, particularly when drawing up to a kerb.

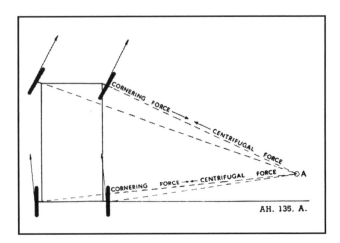

Fig. 6. Slip when cornering causes increased tyre wear.

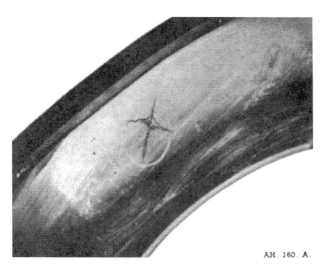

AH. 160. A.

Fig. 7. Severe impact has fractured the casing.

"Spotty Wear"

Fig. 9 shows a type of irregular wear which sometimes develops on front tyres and particularly on near-side front tyres.

The nature of "spotty" wear—the pattern being much worn and little worn at irregular spacings round the circumference—indicates an alternating "slip grip" phenomenon, but it is seldom possible to associate its origin and development with any single cause. There is evidence of camber wear, misalignment, underinflation, or braking troubles.

It is preferable to check all points which may be contributory factors. The front tyres and wheel assemblies may then be interchanged, which will also reverse their direction of rotation, or better still the front tyres may be interchanged with the rear tyres.

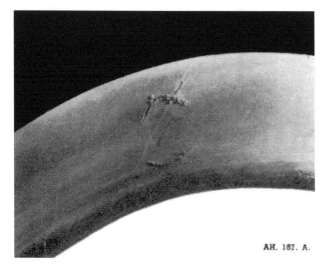

AH. 157. A.

Fig 8. A double fracture caused by the tyre being crushed between the rim and an obstacle, such as the edge of a kerb.

Points for checking are:—
(*a*) Inflation pressures and the consistency with which the pressures are maintained.
(*b*) Brake freedom and balance, shoe settings, lining condition, drum condition and truth.
(*c*) Wheel alignment.
(*d*) Camber and similarity of camber of the front wheels.
(*e*) Play in hub bearings, swivel pin bearings, suspension bearings, and steering joints.

AH. 159. A.

Fig. 9. Irregular "spotty" wear, to which a variety of causes may contribute.

(*f*) Wheel concentricity at the tyre bead seats.
(*g*) Balance of the wheel and tyre assemblies.
(*h*) Conditions of road springs and shock absorbers.

Corrections which may follow a check of these points will not always effect a complete cure and it may be necessary to continue to interchange wheel positions and reverse directions of rotation at suitable intervals.

Irregular wear may be inherent in the local road conditions such as from a combination of steep camber, abrasive surfaces, and frequent hills and bends. Driving methods may also be involved. Irregular wear is likely to be more prevalent in summer than in winter, particularly on new or little worn tyres.

Fig. 10. Fins or feathers caused by severe misalignment. With minor misalignment, probably aggravated by road camber, the ribs may have sharp edges instead of upstanding fins. These conditions will usually be accompanied by heel and toe wear across the tread due to its being distorted and worn away laterally instead of in a true rolling direction.

Wheel Alignment and Road Camber

It is very important that correct wheel alignment should be maintained. Misalignment causes a tyre tread to be scrubbed off laterally because the natural direction of the wheel differs from that of the car.

An upstanding fin on the edge of each pattern rib is a sure sign of misalignment and it is possible to determine from the position of the "fins" whether the wheels are toed in or toed out, see fig. 10. Fins on the inside edges of the pattern ribs—nearest to the car—and particularly on the off-side tyre, indicate toe-out.

With minor misalignment the evidence is less noticeable and sharp pattern edges may be caused by road camber even when wheel alignment is correct. In such cases it is better to make sure by checking with an alignment gauge.

Road camber affects the direction of the car by imposing a side thrust and if left to follow its natural course the car will drift to the near side. This is instinctively

corrected by steering towards the road centre. As a result the car runs crab-wise. Fig. 11 shows, in exaggerated form, the effect this has upon the tyres.

The near front tyre sometimes persists in wearing faster and more unevenly than the other tyres even when the mechanical condition of the car and tyre maintenance are satisfactory. The more severe the average road camber the more marked will this tendency be. This is an additional reason for the regular interchange of tyres.

Camber Angle

This angle normally requires no attention unless disturbed by a severe impact, however, it is always advisable to check this angle if steering irregularities develop, see p. J/7.

Wheel camber usually combined with road camber, causes a wheel to try to turn in the direction of lean, due to one side of the tread attempting to make more revolutions per mile than the other side. The resulting increased tread shuffle on the road and the off centre tyre loading tend to cause rapid and one sided wear. If wheel camber is excessive for any reason the rapid and one sided tyre wear will be correspondingly greater. Unequal cambers introduce unbalanced forces which try to steer the car one way or the other. This must be countered by steering in the opposite direction which results in faster tread wear.

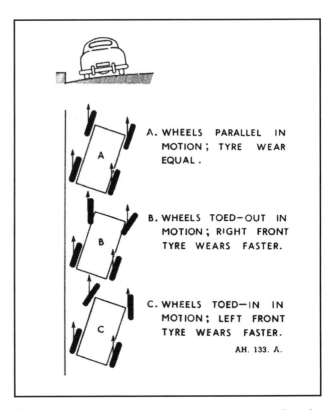

A. WHEELS PARALLEL IN MOTION; TYRE WEAR EQUAL.

B. WHEELS TOED-OUT IN MOTION; RIGHT FRONT TYRE WEARS FASTER.

C. WHEELS TOED-IN IN MOTION; LEFT FRONT TYRE WEARS FASTER.

AH. 133. A.

Fig. 11. Exaggerated diagram of the way in which road camber affects a car's progress.

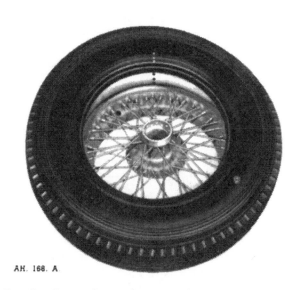

AH. 168. A.

Fig. 12. Correct fitting relationship of Dunlop covers and tubes.

TYRE AND WHEEL BALANCE

Static Balance

In the interests of smooth riding, precise steering and the avoidance of high speed "tramp" or "wheel hop," all Dunlop tyres are balance checked to predetermined limits.

To ensure the best degree of tyre balance the covers are marked with white spots on one bead and these indicate the lightest part of the cover. Tubes are marked on the base with black spots at the heaviest point. By fitting the tyre so that the marks on the cover bead exactly co-incide with the marks on the tube a high degree of tyre balance is achieved, see fig. 12.

When using tubes which do not have coloured spots it is usually advantageous to fit the covers so that the white spots are at the valve position.

Some tyres are slightly outside standard balance limits and are corrected before issue by attaching special loaded patches to the inside of the covers at the crown. These patches contain no fabric, they do not affect the local stiffness of the tyre and should not be mistaken for repair patches—they are embossed "Balance Adjustment Rubber".

The original degree of balance is not necessarily maintained and it may be affected by uneven tread wear, by cover and tube repairs, by tyre removal and refitting or by wheel damage and eccentricity. The car may also become more sensitive to unbalance due to normal wear of moving parts.

Should roughness or high speed steering troubles develop, and mechanical investigation fails to disclose a possible cause, wheel and tyre balance should be suspected.

A tyre balancing machine is marketed by the Dunlop Company to enable service stations to deal with such cases.

Dynamic Balance

Static unbalance, as its name implies, can be measured when the tyre and wheel assembly is stationary. There is, however, another form known as dynamic unbalance which can be detected only when the assembly is revolving.

There may be no heavy spot—that is, there may be no natural tendency for the assembly to rotate about its centre due to gravity—but the weight may be unevenly distributed each side of the tyre centre line, see fig. 13.

Laterally eccentric wheels give the same effect. During rotation the offset weight distribution sets up a rotating couple which tends to steer the wheel to right and left alternately.

Dynamic unbalance of tyre and wheel assemblies can be measured on the Dunlop tyre balancing machine and suitable corrections made when a car shows sensitivity to this form of unbalance. Where it is clear that a damaged wheel is the primary cause of severe unbalance it is advisable for the wheel to be replaced.

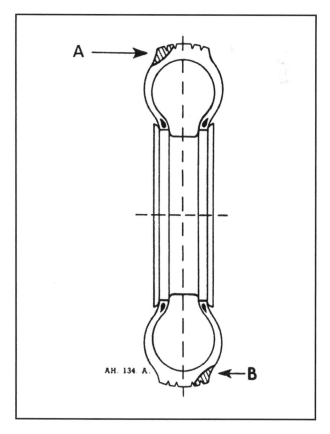

AH. 134. A.

Fig. 13. Dynamic or couple unbalance. Produces wear at 'A' and 'B'

Changing Position of Tyres

Reference has already been made to irregular tread wear which is confined almost entirely to front tyres and there may be different rates of wear between one tyre and another.

It is, therefore, recommended that front tyres be interchanged with rear tyres at least every 2,000 miles. Diagonal interchanging between near-side front and off-side rear and between off-side front and near-side rear provides the most satisfactory first change because it reverses the direction of rotation.

Subsequent interchanging of front and rear tyres should be as indicated by the appearance of the tyres, with the object of keeping the wear of all tyres even and uniform.

Wheels

The Austin-Healey "100" is fitted with $4J \times 15$ centre lock 42W shell wire wheels. No special maintenance is required with wheels of this type under normal conditions. Should they require any attention, however, they should only be handled by a specialist wheel builder, or referred to the Service Department of the Austin Motor Co. Ltd., or the nearest Dunlop Depot.

Wheel Wobble: The lateral variation measured on the vertical inside face of a flange should not exceed $\frac{3}{32}$ in. (2.3812 mm.).

Wheel Lift: On a truly mounted and revolving wheel the difference between the high and low points, measured at any location on either tyre bead seat, should not exceed $\frac{3}{32}$ in. (2.3812 mm.).

Radial and lateral eccentricity outside these limits contribute to static and dynamic unbalance respectively. Severe radial eccentricity also imposes intermittent loading on the tyre. Static balancing does not correct this condition which can be an aggravating factor in the development of irregular wear.

A wheel which is eccentric laterally will cause the tyre to snake on the road, but this in itself has no effect on the rate of tread wear.

At the same time undue lateral eccentricity is undesirable and it affects dynamic balance.

Rim seatings and flanges in contact with the tyre beads should be free from rust and dirt.

Wheel Changing

First loosen the "knock-on" hub cap, then jack up the car. If it is a front wheel which is to be changed the lip on the platform of the screw type jack must project into the recess in the spring plate, whilst the platform should be across the outer rim of the spring plate, the flat **end** between the lower wishbone links.

For lifting the rear wheels, place the lifting platform across the lowest spring leaf, to the rear of the axle, with the lipped end on the outside of the spring and up against the spring "U" bolt, this avoids any turning movement.

After jacking, the hub cap can be screwed right off. The wheel is then pulled off the splined hub.

Refitting the wheel is simply a reversal of this removal procedure, but the splines of the hub and wheel are so fine that the operator should be careful lest in his haste he jams the splines. A little grease should be smeared upon the splines and cone faces of the hub and wheel before refitting. The hub cap threads will also benefit from an occasional application of grease.

Remember that hub caps fitted to right-hand side hubs have left-hand threads, left-hand hubs have right-hand threads, however, the direction for turning is clearly marked on each cap. Caps should be finally tightened with a mallet.

Tyre Removal

(1) Remove all valve parts to deflate the tyre and push both tyre beads off the rim seats.

(2) Commence to remove the bead on the valve side of the cover. Insert a lever at the valve position and, while pulling on this lever, push the bead into the well of the rim diametrically opposite the valve.

(3) Insert a second lever about 2 in. away from the first lever and gradually prise the bead over the rim flange.

(4) Continue with one lever while holding the removed portion of the bead with the other lever. The tube can then be removed.

(5) Stand the cover upright with the wheel in front.

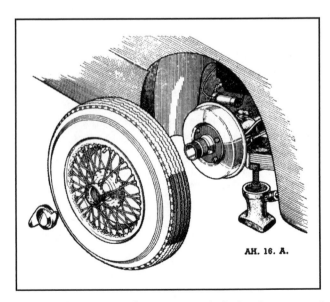

Fig. 14. *Showing the jack in position and wheel and cap removed from the hub.*

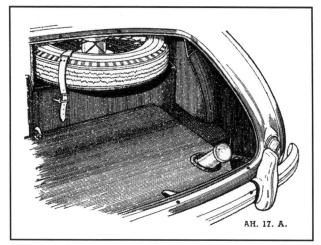

Fig. 15. *Stowage of the spare wheel in the luggage compartment.*

(6) Insert a lever from the front between the bead and the flange and pull the cover back over the flange.

(7) If difficult to remove, keep the strain on the bead with the lever and tap off with a rubber mallet.

Tyre Replacement

(1) Place the cover on top of the wheel and push as much as possible of the lower bead by hand into the well of the rim. Insert a lever to prise the remaining portion of the lower bead over the rim flanges.

(2) Slightly inflate the tube until it begins to round out and insert it in the cover with the valve through the hole in the rim. Take care that the valve, which is fitted in the side of the tube, is on the correct side of the rim and that the tube and spot markings coincide; a point of balance already described.

(3) Commence to fit the second bead by pushing it into the well of the rim diametrically opposite the valve.

(4) Lever the bead over the flange either side of this position, finishing at the valve, when the bead will be completely fitted.

(5) Ease the valve in the rim hole and push upwards by hand to enable the beads to seat correctly and then pull the valve firmly back into position.

(6) Inflate the tyre and see that the beads are seated evenly round the rim: check by the line on the cover.

Note: Water on levers considerably eases the fitting and removing of beads.

SERVICE JOURNAL REFERENCE

NUMBER	DATE	SUBJECT	CHANGES

ELECTRICAL EQUIPMENT

THE electrical equipment is designed to give long periods of service without need for adjustment or cleaning. The small amount of attention which is required is described under "Lubrication and General Maintenance".

Under "General Information" details are given on the operation of the various items of the equipment and descriptions on the method of setting the lamp beams and fitting replacements, such as bulbs, high tension cables, bearing bushes, etc. which may become necessary from time to time.

GENERAL DATA

Battery

Two series-connected 6-volt units, SLTW11E. 50 ampere-hour capacity at 10-hour rate.

Dynamo

C45 PV-5. Maximum output 22 amperes at 13.5 volts.

Distributor

DM2P4. Fitted with Automatic Advance Units. Centrifugal advance begins at 300–500 distributor r.p.m. with maximum advance of 16°–18° at 2000 r.p.m. Vacuum advance begins at 6–8 in. Hg with 6° advance at 9–11½ in. Hg. Firing angles: 0°, 90°, 180°, 270°, ±1°. Closed period of contacts: 60°±3°. Open period of contacts: 30°±3°. Contact breaker gap: 0.014 in. to 0.016 in. Contact breaker spring tension, measured at contacts: 20–24 oz. Capacitor: 0.2 microfarad.

Ignition Coil

B12/1. Resistance of Primary Winding 4.0–4.4 ohms.

Control Box

RB106/1. Compensated voltage control regulator and cut-out unit.

Fuse Unit

SF6. 50-ampere fuse connected between terminals A1–A2 and 35-ampere fuse connected between terminals A3–A4. Two spare fuses.

Starting Motor

M418G. Lock Torque 17 lb./ft. with 440–460 amperes and 7.4–7.0 volts. Light load running current 45 amperes at 12 volts.

Solenoid Starting Switch

ST950. Energised by Starting Push SS5.

Direction Indicators

Either Flasher Unit FL2 (on early models) or Flasher Unit FL3 (on later models) together with model DB10 relay, WL11 warning light and double-filament sidelamp bulbs. In the event of simultaneous operation of flasher unit and brake light switch, responses to the flasher unit will over-ride the brake light.

Battery Master Switch

ST330. When operated, isolates battery from chassis and short-circuits contact breaker.

Horns

HF1748. Matched high-note and low-note high frequency horns.

Windscreen Wiper

CRT15. Built-in thermostatic overload protection. Cuts out at 90°–95°C. Cuts in at 60°C. Normal current of motor 2.0–3.25 amperes. Field resistance 8.4–9.0 ohms.

Overdrive Components

Switch 2TS; Two Relays SB40/1; Centrifugal Switch OCS1; Throttle Switch RTS1; Top Gear Switch SS10/1; Solenoid TGS1; Warning Light WL i1.

Lamps

Headlamps F700 Mk. VI; Sidelamps, Stop-tail, Front and Rear Flashers 488; Number Plate Lamp 467/2.

Bulbs

Headlamps: Right-hand drive cars, Lucas No. 354 12 volt 42/36 watt Prefocus.

Left-hand drive cars, Lucas No. 301 12 volt 36/36 watt Prefocus, or 42/36 watt No. 355.

RH and LH cars (Europe), Lucas No. 370 12 volt 45/40 watt Prefocus or No. 350, 35/35 watt.

Sidelamps, Stop-Tail Lamps and Flasher Lights, Lucas No. 361 12 volt 6/18 watt.

Number Plate Lamp, Lucas No. 222 12 volt 4 watt.

Panel and Warning Lights, Lucas No. 987 12 volt 2.2 watt.

Miscellaneous Switches

Panel 10A; Lighting PPG1; Wiper PS7/2; Ignition S45; Steering Column Control CC1; Dipper FS22/1.

LUBRICATION AND GENERAL MAINTENANCE

FIRST 500 MILES
(800 km.)

Distributor

Remove the moulded distributor cap and turn the engine over by engaging top gear and rolling the car forward until the contacts in the distributor are fully opened. Check the gap with a gauge of thickness 0.014–0.016 in. If the setting is correct the gauge should be a sliding fit. Should the gap vary appreciably from the gauge, the contact breaker must be adjusted. To carry out the adjustment, keep the engine in the position to give the maximum opening of the contacts and with the screwdriver supplied in the tool kit, slacken the two screws which secure the contact plate. Move the plate until the gap is set to the thickness of the gauge and then fully tighten the locking screws. Recheck the gap.

EVERY 1,000 MILES
(1,600 km.)

Battery

About every 1,000 miles, or more often in warmer climates, remove the vent plugs from the top of each of the battery cells and examine the level of the electrolyte. If necessary, add distilled water until the top edges of the separators are just covered.

Do not fill above this level, otherwise the excess electrolyte will be spilled from the cell. A hydrometer will be found useful for topping up, as it prevents distilled water being spilled on the top of the battery.

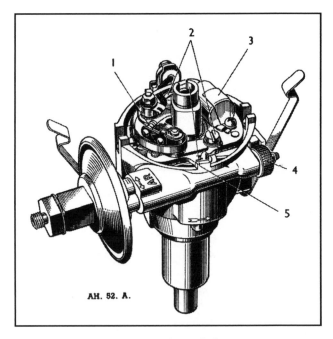

Fig. 2. Distributor platform.
1. Contact points. 2. Contact adjusting screws. 3. Capacitor.
4. Micrometer adjuster. 5. Cam and drive shaft oiling points.

In very cold weather it is essential that the car should be used immediately after topping up the battery, this ensures that the distilled water is thoroughly mixed with the electrolyte. Neglect of this precaution may result in the distilled water freezing and so causing damage to the battery.

When examining the cells, do not hold naked lights near the vent holes as there is a danger of igniting the gas coming from the plates.

EVERY 3,000 MILES
(4,800 km.)

Carry out the procedure prescribed for 1,000 miles together with the following.

Distributor Lubrication

Cam: Lightly smear the cam with a very small amount of clean engine oil, or Mobilgrease No. 2.

Cam Bearing and Distributor Shaft: Pull the rotor arm from the top of the spindle and add a few drops of thin machine oil to lubricate the cam bearing and distributor shaft. Do not remove the screw exposed to view, as this screw is drilled to enable the oil to pass through it. Take care to fit the rotor arm correctly, pushing it on the shaft as far as it will go.

Automatic Timing Control: Add a few drops of thin machine oil through the aperture at the edge of the

AH. 15. A.

Fig. 1. Access to the batteries is gained through panel behind the seats.

contact breaker. **Do not allow any oil to get on or near the contacts.**

Contact Breaker Pivot: Lightly lubricate the pivot on which the contact breaker lever works. Do not allow oil or grease to get on to the contacts.

EVERY 6,000 MILES
(9,600 km.)

Carry out the procedure for every 1,000 and 3,000 miles together with the following.

Distributor—Cleaning

Wipe the inside and outside of the moulding with a soft dry cloth, paying particular attention to the spaces between the metal electrodes. See that the small carbon brush on the inside of the moulding moves freely in its holder.

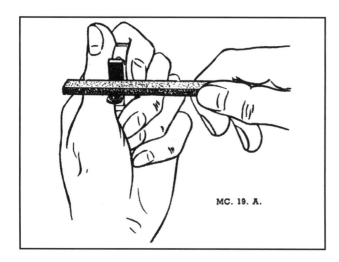

Fig. 3 Trueing the contacts with the aid of a carborundum stone.

Examine the contact breaker. The contacts must be free from grease and oil. If they are burned or blackened, clean them with a fine carborundum stone or very fine emery cloth, afterwards wiping away any trace of dirt or metal dust with a petrol-moistened cloth. Cleaning of the contacts is made easier if the contact breaker lever carrying the moving contact is removed. To do this, remove the nut, washer, insulating piece and connections from the post to which the end of the contact breaker spring is anchored. Before refitting the contact breaker, smear the pivot with clean engine oil or Mobilgrease No. 2.

After cleaning the contacts check the contact breaker setting and if necessary make adjustments as described for the first 500 miles.

High Tension Cables

Examine the high tension cables. Any of which have the insulation cracked or perished, or show signs of

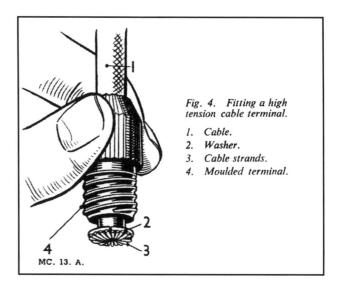

Fig. 4. Fitting a high tension cable terminal.

1. Cable.
2. Washer.
3. Cable strands.
4. Moulded terminal.

MC. 13. A.

damage in any other form, must be replaced by 7 mm. rubber-covered ignition cable. The method of connecting high tension cables to the coil and distributor is to thread the knurled moulded nut over the cable, bare the end for about $\frac{1}{4}$ in., thread the wire through the washer removed from the original cable, and bend back the wire strands. Screw the nut into its terminal, see Fig. 4.

EVERY 9,000—12,000 MILES
(14,400—19,200 km.)

Dynamo—Lubrication

Early Models: Repack the dynamo lubricator with H.M.P. grease at every 9,000 to 12,000 miles running. Ensure that the felt pad and spring are replaced before screwing the lubricator in position. See Fig. 5.

Later Models: Inject high quality medium viscosity engine oil (SAE 30) into the oiling hole in the commutator end bracket bearing housing. See illustration.

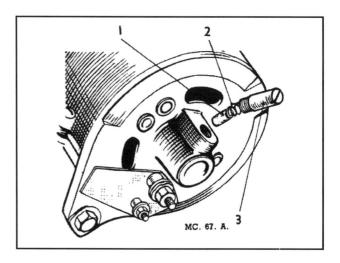

Fig. 5. Dynamo lubrication.

1. *Wick lubricator.*   2. *Spring.*   3. *Screwed plug.*

GENERAL INFORMATION

BATTERY

Occasionally examine the condition of the battery by taking hydrometer readings; there is no better method of ascertaining the state of charge of the battery. The hydrometer contains a graduated float which indicates the specific gravity of the acid in the cell from which the sample is taken.

The specific gravity readings and their indications are as follows:—

Reading	State of Battery
1.280–1.300	Battery fully charged
Approx. 1.210	Battery half charged
Below 1.150	Battery fully discharged

These figures are given assuming an electrolyte temperature of 60°F. If the electrolyte temperature exceeds this, .002 must be added to hydrometer readings for each 5°F. rise to give the true specific gravity at 60°F. Similarly, .002 must be subtracted from hydrometer readings for every 5°F. below 60°F.

The readings for each of the cells should be approximately the same. If one cell gives a reading very different from the rest, it may be that the electrolyte has been spilled or has leaked from one of the cells, or there may be an internal fault. In the latter case it is advisable to have the battery examined by a battery specialist.

Should the battery be in a low state of charge, it should be recharged by taking the car for a long daytime run or by charging from an external source of D.C. supply, at a current rate of 5 amps until the cells are gassing freely.

After examining the battery, check the vent plugs, making sure that the air passages are clear, and screw the plugs into position. Wipe the top of the battery to remove all dirt and moisture.

Storage

If a battery is to be out of use for any length of time, it should first be fully charged and then given a freshening charge about every fortnight.

A battery must never be allowed to remain in a discharged condition, as this will cause the plates to become sulphated.

Initial Filling and Charging

The battery will usually be supplied filled and initially charged, or, for export markets, "dry-charged."

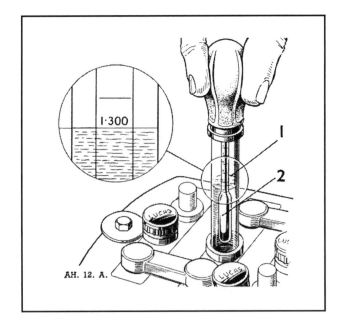

Fig. 6. *Using a hydrometer to test the specific gravity of the battery acid.*   *1. Reading.*   *2. Float.*

However, if it should be necessary to prepare for service a new battery, supplied dry and *uncharged*, proceed as follows:—

(a) Preparation of Electrolyte: The specific gravity of the electrolyte necessary to fill the new battery, and the specific gravity at the end of the charge, are as follows:—

Climate	(Corrected to 60°F.)	
	S.G. of Filling Acid	S.G. at End of Charge
Below 80°F. . . .	1.350	1.280–1.300
Between 80–100°F. .	1.320	1.250–1.270
Over 100°F. . . .	1.300	1.220–1.240

The electrolyte is prepared by mixing distilled water and concentrated sulphuric acid of 1.835 S.G. The mixing must be carried out in a lead-lined tank or a suitable glass or earthenware vessel. Steel or iron containers must *not* be used. The acid must be added slowly to the water, while the mixture is stirred with a glass rod. *Never add the water to the acid*, as the resulting chemical reaction may have dangerous consequences. To produce electrolyte of the correct specific gravity as previously stated, use proportions of acid and distilled water as follows:—

To obtain Specific Gravity (corrected to 60°F.)	Add 1 part by volume of 1.835 S.G. acid to distilled water by volume as below:
1.350	1.8 parts
1.320	2.2 parts
1.300	2.5 parts

Heat is produced by the mixture of acid and water, and it should, therefore, be allowed to cool before pouring it into the battery, otherwise the plates, separators and moulded container may become damaged.

(b) Filling-In and Soaking: The temperature of the filling-in acid, battery and charging room should be above 32°F.

Carefully break the seals in the filling holes and half fill each cell in the battery with diluted sulphuric acid solution of the appropriate specific gravity (according to temperature). The quantity of electrolyte to half fill a two-volt cell is ½ pint.

Allow the battery to stand for at least six hours, in order to dissipate the heat produced by the chemical action of the acid on the plates and separators. Then add sufficient electrolyte to fill each cell to the top of the separators. Allow to stand for a further two hours and then proceed with initial charge.

(c) Duration and Rate of Initial Charge: Charge at a constant current of 3.5 amps until voltage and temperature corrected specific gravity readings show no increase over five successive hourly readings. This period is dependent upon the length of time the battery has been stored since manufacture, and will be from forty to eighty hours, but usually not more than sixty.

Throughout the charge, the acid must be kept level with the tops of the separators in each cell by the addition of acid solution of the same specific gravity as the original filling-in acid.

If during charge, the temperature of the acid in any cell of the battery reaches the maximum permissible temperature of 120°F., the charge must be interrupted and the battery temperature allowed to fall at least 10°F. before charging is resumed.

At the end of the first charge, i.e. when specific gravity and voltage measurements remain substantially constant, carefully check the specific gravity in each cell to ensure that it lies within the limits specified. If any cell requires adjustment, some of the electrolyte must be siphoned off, and replaced with either distilled water or

acid of the strength used for the original filling-in, according to whether the specific gravity is too high or too low respectively. After such adjustment, the gassing charge should be continued for one or two hours to ensure adequate mixing of the electrolyte. Recheck, if necessary, repeating the procedure until the desired result is obtained.

Finally allow the battery to cool and siphon off any electrolyte above the separators.

Preparing "Dry-Charged" Batteries for Service

These batteries are very similar, as far as their operation and external appearance are concerned, to the normal lead-acid type. The difference lies in the fact that these batteries are "dry-charged" and sealed before leaving the factory, so that when they are required for service it is only necessary to fill each cell with sulphuric acid of the correct specific gravity. This procedure also ensures that there is no deterioration of the efficiency of the battery during the storage period before the battery is required for use.

(a) Preparation of Electrolyte: The electrolyte is prepared by mixing together distilled water and concentrated sulphuric acid, usually of S.G. 1.835. This mixing must be carried out in a lead-lined tank or a glass or earthenware vessel. The acid must be added slowly to the water while the mixture is stirred with a glass rod.

Electrolyte of specific gravity 1.270 can be prepared by adding 1 part (by volume) of 1.835 S.G. sulphuric acid to 2.8 parts of distilled water. 1.210 requires 1 part of acid to 4 parts of distilled water.

Heat is produced by the mixture of acid and water, and the electrolyte should be allowed to cool before pouring it into the battery.

(b) Filling the cells: Carefully break the seals in the cell filling holes and fill each cell with electrolyte to the top of the separators, *in one operation.* The temperature of the filling room, battery and electrolyte should be maintained between 60°F. and 100°F. If the battery has been stored in a cool place, it should be allowed to warm up to room temperature before filling.

(c) Batteries filled in this way are 90 per cent. charged and capable of giving a starting discharge one hour after filling. When time permits, however, a short freshening charge will ensure that the battery is fully charged. Such a freshening charge should last for no more than 4 hours, at the normal recharge rate of the battery: 5 amperes.

During the charge the electrolyte must be kept level with the top edge of the separators by the addition of distilled water. Check the specific gravity of the acid at the end of the charge; if 1.275 acid was used to fill the battery, the specific gravity should now be between 1.280 and 1.300; if 1.215, between 1.220 and 1.240.

| Battery | Voltage | A.H. Capacity | | Quantity of electrolyte required | Specific gravity of filling electrolyte | |
		at 10 hr. rate	at 20 hr. rate		Climates not normally above 90°F. (32°C.)	Climates frequently above 90°F.
SLTZ11E	6 per unit	50	57	3 pints per 6-volt unit	1.270	1.210

DYNAMO

In the event of a fault occurring in the charging circuit proceed as follows:—

(i) Inspect the dynamo driving belt and if necessary adjust by turning the dynamo on its mounting to take up any undue slackness. Care should be taken to avoid over-tightening the belt and to see that the machine is properly aligned, otherwise undue strain will be thrown on the dynamo bearings.

(ii) Check that the large dynamo terminal is connected to control box terminal "D" and that the small dynamo terminal is connected to control box terminal "F". Check control box earth connection "E" for tightness.

(iii) Switch off all lights and accessories. Disconnect the cables from the dynamo terminals and connect the terminals with a short length of wire.

(iv) Start the engine and set to run at normal idling speed.

(v) Clip the negative lead of a moving coil 0–20 voltmeter to one dynamo terminal and the positive lead to a *good* earthing point on the dynamo yoke.

(vi) Gradually increase the engine speed. The voltmeter reading should rise rapidly and without fluctuation. Do not allow the voltmeter reading to reach 20 volts and do not race the engine in order to increase the voltage. It is sufficient to run the dynamo up to 1000 r.p.m. If no reading is obtained, the brushgear may require checking. A reading of 0.5–1.0 volt may indicate a field winding fault whilst a reading of 4.0–5.0 volts may indicate a faulty armature.

Removing and Replacing the Dynamo

To remove the dynamo, slacken the setpin which secures the adjusting link to the dynamo casing, remove the nut and washer from the adjusting stud, and release the two nuts and bolts holding the dynamo to its mounting bracket.

Replacement is a reversal of the removal procedure, with the adjustment nut being tightened first, once the dynamo position has been finalised.

Dismantling the Dynamo

Remove the dynamo from its mounting bracket and transfer it to a clean work bench.

Take off the dynamo pulley by releasing the securing nut and washer. If the pulley is exceptionally tight on the shaft and woodruff key, use a suitable extractor such as tool 18G 2. Remove the cover band, hold back the brush springs and remove the brushes from their holders.

Remove the nut and washers from the smaller dynamo terminal. Unscrew and withdraw the two through bolts. Separate the commutator end bracket from the dynamo yoke. Withdraw the drive end bracket, together with the armature, from the dynamo yoke.

The drive end bracket need not be separated from the shaft unless the bearing is suspect and requires examination, or if the armature is to be replaced. In such cases the armature should be removed from the end bracket using a hand press.

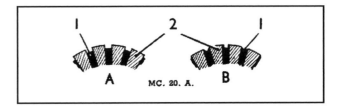

Fig. 7. *Undercutting the commutator. 'A' is the correct and 'B' the incorrect method. 1. Mica insulation. 2. Segments.*

Commutator

Examine the commutator and if burned or blackened, clean with a petrol-moistened cloth, or in bad cases by carefully polishing with very fine glass paper, if necessary, undercut the insulation to a depth of $\frac{1}{32}$ in. with a hacksaw blade ground down to the thickness of the insulation.

Armature

Check the armature by means of a growler test and volt drop test, and make proof as to the condition of the insulation by connecting a test lamp, at mains voltage, between the commutator segments and the shaft.

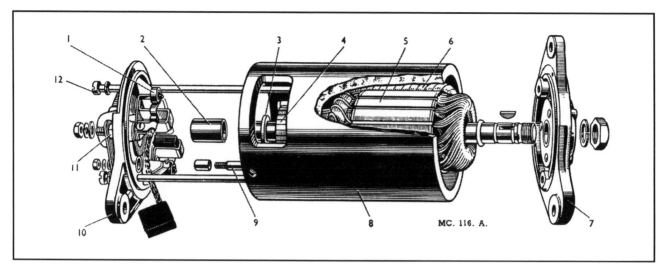

Fig. 8. An exploded view of the dynamo.

1. Brush spring
2. Commutator end bearing.
3. Thrust collar.
4. Commutator.
5. Armature
6. Field coil.
7. Driving end bracket.
8. Yoke.
9. Field terminal.
10. Commutator end bracket.
11. Terminal.
12. Through bolt.

Brushgear

Examine the brushes. If they are worn so that they do not make good contact on the commutator or if the brush flexible is exposed on the running face, take out the screw securing the eyelet on the end of the brush flexibles and remove the brushes. Fit new brushes into their holders and refit the eyelet screws. After fitting, the new brushes must be bedded to the commutator.

Field Coils

Test the resistance of the field coils by means of an ohmmeter. The reading on the ohmmeter should be 6 ohms. If such an instrument is not available, connect a 12-volt D.C. supply with an ammeter in series, between the field terminal and the dynamo frame. The ammeter reading should be approximately 2 amps. If there is no reading, the field coils are open circuited and must be replaced.

To test for earthed field coils, unsolder the end of the field winding from the earth terminal on the dynamo yoke, and with a test lamp connected to the supply mains, check between the field terminal and dynamo yoke. If the lamp lights, the field coils are earthed and must therefore be replaced.

When replacing field coils, an expander should be used so as to press the pole shoes into position. A few taps on the outside of the dynamo frame with a copper-faced mallet will assist the expander to seat the pole shoes. When the pole shoes are finally home, fully tighten up the fixing screws and caulk, to lock them in position.

Bearings

Bearings which are worn to such an extent that they will allow excessive side movement of the armature shaft must be replaced.

Commutator End

To remove and replace the bearing bush at the commutator end proceed as follows:—

(a) Insert an $\frac{11}{16}$ in. tap squarely into the open end of the defective bush and withdraw the bush.

(b) The order of bearing reassembly with later dynamos not fitted with screw type lubricator is:—

1. Felt Ring.
2. Aluminium Disc.
3. Bearing Bush.

The bearing should be pressed into the bearing housing until it is flush with the inner face of the end bracket.

Note: Before fitting a new porous bronze bearing bush, it should be immersed for 24 hours in clean thin engine oil.

Driving End

The ball bearing at the driving end is replaced as follows:—

(a) Knock out the rivets which secure the bearing retaining plate to the end bracket and remove the plate.

(b) Press the bearing out of the end bracket and remove the corrugated washer, felt washer and oil retaining washer.

(c) Before fitting the replacement bearing see that it is clean and lightly pack it with high melting point grease.

(d) Place the oil retaining washer, felt washer and corrugated washer in the bearing housing in the end bracket.

(e) Locate the bearing in the housing and press it home. The outer bearing journal is a light push fit in the bearing housing.

Note: When fitting a drive end bracket to the armature shaft, the inner journal of the bearing *must* be supported by a mild steel tube. This tube should be approximately 4 in. long and $\frac{1}{8}$ in. thick with an internal diameter of $\frac{11}{16}$ in.

Do not use the drive end bracket as a support for the bearing when fitting an armature.

(f) Fit the bearing retaining plate. Insert new rivets from the inside of the end bracket and open the rivets by means of a punch to secure the plate rigidly in position.

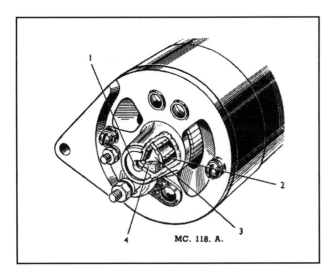

Fig. 9. Dynamo lubrication (later type).
1. Oil hole.   2. Bronze brush.   3. Aluminium disc.   4. Felt ring.

Reassembling the Dynamo

In the main, reassembly of the dynamo is a reversal of the dismantling procedure. Before refitting the dynamo, however, fill the lubricator with H.M.P. grease, or inject oil into the C.E. bracket, as previously described.

STARTER

Normal Service

If difficulty is experienced with the starter not meshing correctly with the flywheel, it may be that the starter drive requires cleaning. The pinion should move freely on the screwed sleeve, if there is any dirt or foreign matter on the sleeve, it must be washed with paraffin.

In the event of the starter pinion becoming jammed in mesh with the flywheel, it can usually be freed by turning the starter armature by means of a spanner applied to the shaft extension at the commutator end. This is accessible when the cap protecting the armature shaft is removed.

Removing the Starter

This operation is achieved by releasing the main cable from the terminal at the front of the starter, and by removing the engine to gearbox nuts and bolts holding the starter body in position.

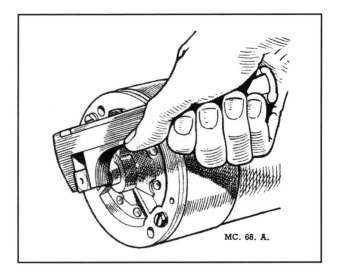

Fig. 10. Using a spanner to rotate the armature to free a pinion jammed in mesh with the flywheel.

Dismantling the Starter

Starter Drive: Remove the cotter pin from the nut at the end of the shaft and then unscrew the square shaft nut. Release the main spring, buffer washer, screwed sleeve and pinion. Finally withdraw the restraining spring, sleeve and collar.

Starter Motor: Take off the cover band at the commutator end, hold back the brush springs and take out the brushes from their holders. Extract the two through-bolts when the armature complete with driving end bracket may be withdrawn. Remove the terminal nuts and washers from the terminal post on the commutator end bracket and withdraw the bracket from the starter yoke.

Commutator

Examine the commutator and if burned or blackened clean with a petrol-moistened cloth, or in bad cases, by carefully polishing with very fine glass paper. **Do not undercut the insulation.**

Armature

Examination of the armature will in most cases reveal the cause of failure. An example of such a failure is when the conductors are lifted from the commutator, due to the starter being engaged while the engine is running and so causing the armature to be rotated at an excessive speed. A damaged armature must in all cases be replaced—no attempts should be made to machine the armature core to true a distorted armature shaft.

Brushes

Examine the brushes. If they are worn so that they do not make good contact on the commutator, or if the brush flexibles are exposed on the running face

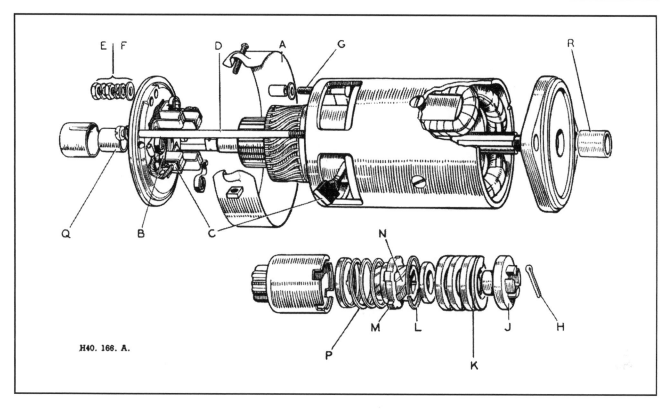

Fig. 11. *The starter in exploded form.*

A.	Cover band.	F.	Terminal washers.	L.	Retaining ring.
B.	Brush spring.	G.	Terminal post.	M.	Control nut.
C.	Brushes.	H.	Split pin.	N.	Sleeve.
D.	Through bolt.	J.	Shaft nut.	P.	Restraining spring.
E.	Terminal nuts.	K.	Retaining spring.	Q.	Bearing bush.

they must be replaced. Two of the brushes are connected to terminal eyelets on the brush boxes whilst the others are connected to tappings on the field coils.

The flexible connectors must be removed by unsoldering and the connectors of the new brushes secured in their places by soldering. The brushes are preformed so that bedding to the commutator is unnecessary.

Field Coils

The field coils can be tested for an open circuit by connecting a 12-volt battery and test lamp between the tapping points on the field coils at which the brushes are connected. If the lamp does not light there is an open circuit in the wiring of the field coils.

Lighting of the lamp does not necessarily mean that the field coils are in order, it is quite feasible that one of them may be earthed to a pole shoe or to the starter yoke. This may be checked with a test lamp connected from the supply mains, the test leads being connected to one of the tapping points on the field coils and to a clean part of the starter yoke. Should the lamp light it indicates that the field coils are earthed and must be replaced.

When replacing field coils the procedure as detailed in the dynamo section should be followed.

Bearings

Bearings which are worn to such an extent that they will allow excessive side play of the armature shaft must

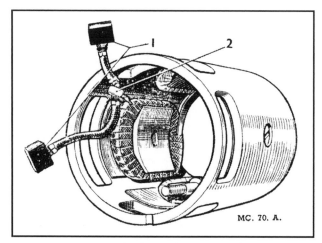

Fig. 12. *The starter yoke showing brushes (1) with attachment (2) at the field coils.*

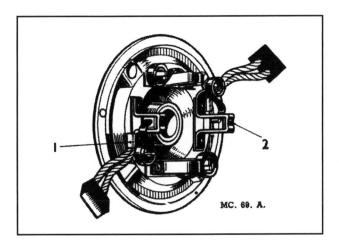

Fig. 13.   *Commutator end bracket.*
1. Brush eyelet fixing.   2. Brush box.

be replaced. To replace the bearing bushes adopt the procedure described for the bushes of the dynamo commutator end.

Reassembling the Starter

The reassembly of the starter is a reversal of the dismantling procedure.

STARTER SWITCH

Testing in Position

Press the starter push and listen for the starter switch to operate. If it does not operate connect a 12-volt supply directly across the small terminal on the solenoid

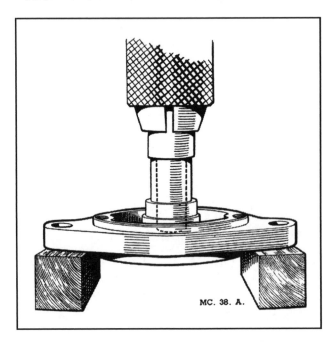

Fig. 14.   *Fitting a new bronze bush to the starter end bracket using a mandrel and press.*

switch and the switch body. Should the switch still fail to operate, a replacement unit must be fitted. If the switch operates but does not complete the circuit to the starter (checked by means of a 12-volt test lamp between starter terminal on switch and earth) an indication is given that the contacts are faulty and the switch must be replaced.

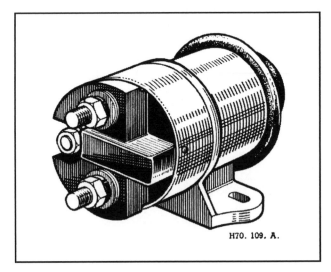

Fig. 15.   *Solenoid starter switch.*

DISTRIBUTOR

Description

The coil ignition equipment comprises a high tension induction coil and a combined distributor, contact breaker and automatic timing control assembly driven at half engine speed via the camshaft. Current flowing through the primary or low tension winding of the coil sets up a strong magnetic field about it. This current is periodically interrupted by a cam-operated contact breaker driven from the engine. The subsequent collapse of the magnetic field induces a high voltage in the secondary winding of the coil. At the same time, a rotor arm in the distributor connects the secondary winding of the coil with one of a number of metal electrodes from which cables lead to the sparking plugs in the engine cylinders. Thus, a spark is arranged to occur in the cylinder under compression at the exact moment required to produce combustion of the mixture.

Mounted on the distributor driving shaft, immediately beneath the contact breaker, is an automatic timing control mechanism. It consists of a pair of spring-loaded governor weights, linked by lever action to the contact breaker cam. At slow engine speeds, the spring force maintains the cam in a position in which the spark is slightly retarded. Under the centrifugal force imparted by high engine speeds, the governor weights swing out against the spring pressure to advance the contact breaker cam, and thereby the spark, to suit engine conditions at the greater speed.

A vacuum-operated timing control is also fitted, designed to give additional advance under part-throttle conditions. The inlet manifold of the engine is in direct communication with one side of a spring-loaded diaphragm. This diaphragm acts through a lever mechanism to rotate the heel of the contact breaker about the cam, thus advancing the spark for part-throttle operating conditions. There is also a micrometer adjustment by means of which fine alterations in timing can be made to allow for changes in running conditions, e.g. state of carbonisation, change of fuel, etc. The combined effects of the centrifugal and vacuum-operated timing controls give added efficiency over the full operating range of the engine, with a corresponding economy in fuel consumption.

A completely sealed metallised paper capacitor is utilised in this distributor model. One feature of this type of capacitor is its property of being self-healing; should the capacitor break down, the metallic film around the point of rupture is vaporised away by the heat of the spark, so preventing a permanent short circuit. Capacitor failure will be found to be most infrequent with this new type.

A measure of radio and television interference suppression is provided by the carbon brush that forms the connection to the rotating electrode of the distributor. This brush, longer than is usual and formed of resistive carbon, has the effect of a suppression resistor in the lead from the coil to the distributor.

Dismantling

When dismantling, carefully note the positions in which the various components are fitted, in order to ensure their correct replacement on reassembly. If the driving dog or gear is offset, or marked in some way for convenience in timing, note the relation between it and the rotor electrode and maintain this relation when reassembling the distributor.

Spring back the securing clips and remove the moulded cover. Lift the rotor arm off the spindle carefully levering with a screwdriver if it is tight.

Slacken and remove the nut securing the end of the contact breaker spring, taking care not to lose the two fibre insulating washers. Lift off the contact breaker lever and the fibre washer beneath it.

Remove the two screws securing the fixed contact plate, and remove the plate. (See note on Replacement Contacts.) Withdraw the single securing screw and remove the capacitor.

Remove the two screws at the edge of the contact breaker base plate.

Remove the split pin securing the vacuum unit link to the rotating contact breaker plate.

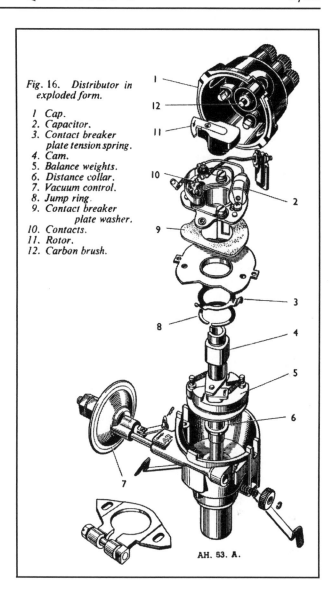

Fig. 16. Distributor in exploded form.

1 Cap.
2. Capacitor.
3. Contact breaker plate tension spring.
4. Cam.
5. Balance weights.
6. Distance collar.
7. Vacuum control.
8. Jump ring.
9. Contact breaker plate washer.
10. Contacts.
11. Rotor.
12. Carbon brush.

AH. 53. A.

Lift the contact breaker base assembly out of the distributor body. Note that the low tension terminal and its moulded block are attached to the lower plate.

The contact breaker base assembly can be dismantled by removing the circlip and star washer located under the lower plate.

Remove the circlip on the end of the micrometer timing screw, and turn the micrometer nut until the screw and the vacuum unit assembly are freed. Take care not to lose the ratchet and coil type springs located under the micrometer nut.

Take out the screw inside the cam and remove the cam and cam foot. The weights, springs and toggles of the automatic timing control can now be lifted off the action plate. Remove the driving dog or gear from the shaft, and lift the shaft out of its bearing.

The single long bearing bush used in this distributor can be pressed out of the shank by means of a shouldered mandrel.

Reassembly

If the bearing has been removed the distributor must be assembled with a new bush fitted. The bush should be prepared for fitting by allowing it to stand completely immersed in thin engine oil for at least 24 hours. In cases of urgency this period of soaking may be shortened to 2 hours by heating the oil to 100°C.

Press the bearing into the shank, using a shouldered, polished mandrel of the same diameter as the shaft. **Under no circumstances should the bushes be overbored by reamering or any other means, since this will impair the porosity and thereby the effective lubricating quality of the bushes.**

Place the distance collar over the shaft, smear the shaft with clean engine oil and fit it into its bearing.

Refit the vacuum unit into its housing and replace the springs, milled adjusting nut and securing circlip.

Reassemble the automatic timing control. See that the springs are not stretched or damaged, and that there is a washer in position under each toggle. Place the cam and cam foot assembly over the shaft, engaging the projections on the cam foot with the toggles, and fit the securing screw.

Moisten the felt pad underneath the rotating contact breaker plate with a drop of thin machine oil. Fit the rotating plate to the contact breaker base plate and secure with the star washer and circlip. Refit the contact breaker base assembly into the distributor body. Engage the link from the vacuum unit with the bearing bush in the rotating plate, and secure with the split pin. Insert the two base plate securing screws, one of which also secures the earthing lead from the contact breaker plate.

Fit the capacitor into position. The eyelet on the contact breaker earthing lead is held under the capacitor fixing screw. Place the fixed contact plate in position and secure lightly with two screws. One plain and one spring washer must be fitted under each of these screws.

Reassemble the terminal screw and eyelets on the fixed contact plate. Take care to position the two insulating washers correctly, so that the eyelets and the screw cannot touch the fixed contact plate. Place the fibre washer over the contact breaker pivot and refit the contact breaker lever and spring. Set the gap to 0.014 in.–0.016 in. and tighten the securing screws of the fixed contact plate.

Finally fit the rotor arm into position, locating the register and pushing it fully home, and refit the moulded distributor cover.

Replacement Contacts

If the contacts are so badly worn that replacement is necessary, they must be renewed as a pair and not individually. The contact gap must be set to 0.014–0.016 in.; after the first 500 miles' running with the new

contacts in position the setting should be checked and the gap reset to 0.014–0.016 in. This procedure allows for the initial "bedding-in" of the heel of the new contact breaker lever.

CONTROL BOX

Description

This unit contains the cut-out and voltage regulator. The regulator controls the dynamo output in accordance with the load on the battery and its state of charge. When the battery is discharged the dynamo gives a high output, so that the battery receives a quick recharge, which brings it back to its normal state in the minimum possible time.

On the other hand if the battery is fully charged, the dynamo is arranged to give only a trickle charge, which is sufficient to keep it in good condition without any possibility of causing damage to the battery by overcharging.

The cut-out takes the form of an automatic switch for connecting and disconnecting the battery and the dynamo. This is necessary because the battery would otherwise discharge through the dynamo when the engine is stopped or running at a low speed.

Regulator—Adjustment

The regulator is carefully set before leaving the works to suit the normal requirements of the standard equipment, and in general it should not be necessary

Fig. 17. Control box.

1. Locknut.   2. Regulator adjusting screw.   3. Locknut.
4. Cut-out adjusting screw.

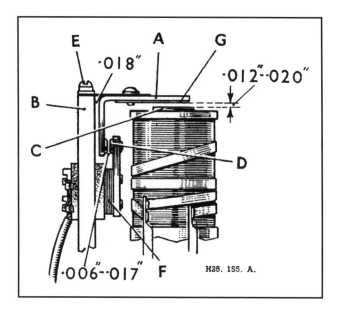

Fig. 18. *Regulator mechanical setting.*

A. *Armature.*   D. *Fixed contact.*
B. *Regulator frame.* E. *Armature fixing screws.*
C. *Bobbin core.*   F. *Packing shims.*

to alter it. If, however, the battery does not keep in a charged condition, or if the dynamo output does not fall when the battery is fully charged, it may be advisable to check the setting and if necessary to readjust.

It is important, before altering the regulator setting, when the battery is in a low state of charge, to check that its condition is not due to a battery defect or to the dynamo belt slipping.

Checking and adjusting the Electrical Setting

The regulator setting can be checked without removing the cover of the control box.

Withdraw the cables from the terminals marked "A" and "A.1" at the control box and join them together. Connect the negative lead of a moving coil voltmeter (0.20 volts full scale reading) to the "D" terminal on the dynamo and connect the other lead from the meter to a convenient chassis earth.

Slowly increase the speed of the engine until the voltmeter needle "flicks" and then steadies; this should occur at a voltmeter reading between the limits given for the appropriate temperature of the regulator.

Setting Temperature	Voltmeter Reading
10 deg. C. (50 deg. F.)	16.1–16.7
20 deg. C. (68 deg. F.)	15.8–16.4
30 deg. C. (86 deg. F.)	15.6–16.2
40 deg. C. (104 deg. F.)	15.3–15.9

If the voltage at which the reading becomes steady occurs outside these limits, the regulator must be adjusted.

Shut off the engine, remove the control box cover, release the locknut (1) Fig. 17 holding the adjusting

screw (2) and turn the screw in a clockwise direction to raise the setting or in an anti-clockwise direction to lower the setting. Turn the adjustment screw a fraction of a turn and then tighten the locknut.

When adjusting, do not run the engine up to more than half throttle, because while the dynamo is on open circuit, it will build up to a high voltage if run at a high speed and in consequence a false voltmeter reading would be obtained.

Mechanical Setting

The mechanical setting of the regulator is accurately adjusted before leaving the works, and provided the armature carrying the moving contact is not removed, the regulator will not require mechanical adjustment. If, however, the armature has been removed from the regulator for any reason, the contacts will have to be reset. To do this proceed as follows:—

(a) Slacken the two armature fixing screws (E) Fig. 18. Insert .018 in. feeler gauge between back of the armature "A" and the regulator frame.

(b) Press back the armature against the regulator frame and down on to the top of the bobbin core with the gauge in position, and lock the armature by tightening the two fixing screws. Remove the 0.018 in. gauge.

(c) Check the gap between the underside of the arm and the top of the bobbin core. This should be .012 in.–.020 in. If the gap is outside these limits correct them by adding or removing shims (F) at the rear of the fixed contact.

(d) Remove the gauge and press the armature down when the gap between the contacts should be .006 in.–.017 in.

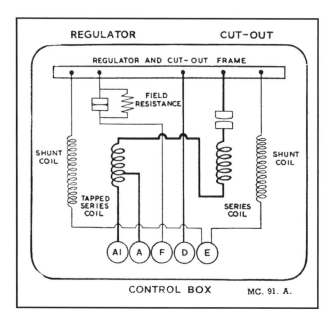

Fig. 19. *Internal connections of control box.*

Cleaning the Contacts

To render the regulator contacts accessible for cleaning, slacken the screws securing the plate carrying the fixed contact. It will be necessary to slacken the upper screw Fig. 18, a little more than the lower screw in order that the contact plate may be swung outwards. Clean the contacts by means of a fine carborundum stone or fine emery cloth. Carefully wipe away all traces of dirt or other foreign matter before finally tightening the securing screws.

Cut-Out Adjustment

If it is suspected that the cutting-in speed of the dynamo is too high, connect a voltmeter between the terminals marked "D" and "E" at the control box and slowly raise the engine speed. When the voltmeter reading rises to about 12.7 to 13.3, the cut-out contacts should close.

If the cut-out has become out of adjustment and operates at a voltage outside these limits it must be reset. To make the adjustment slacken the locknut 3 Fig. 17, turn the adjusting screw 4 a fraction of a turn in a clockwise direction to raise the operating voltage, or in an anti-clockwise direction to lower the voltage. Tighten the locknut after making the adjustment.

Cleaning the Contacts

To clean the contacts insert a strip of fine glass paper between the contacts and then, closing the contacts by hand, draw the paper through. This should be done two or three times, the rough side towards each contact.

IGNITION COIL

The ignition coil requires no attention beyond seeing that the terminal connections are tight and that the exterior is kept clean, particularly between the terminals.

ELECTRIC HORN

All horns, before being passed out of the works, are adjusted to give their best performance, and will give a long period of service without any attention: no subsequent adjustment is required.

If one of the horns fails or becomes uncertain in its action it does not follow that the horn has broken down. First ascertain that the trouble is not due to a loose or broken connection in the wiring of the horn. If both horns fail or become uncertain in action, the trouble is probably due to a blown fuse or discharged battery. If the fuse has blown, examine the wiring for the fault and replace with the spare fuse provided.

It is also possible that the performance of a horn may be upset by the fixing bolts working loose, or by some

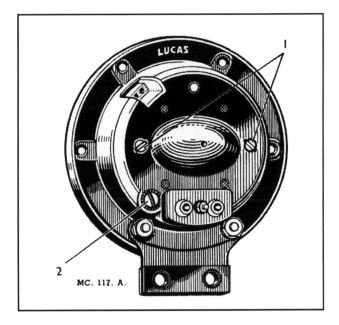

Fig. 20. Rear view of the horn.
1. Coil securing screws.   2. Adjustment screw.

component near the horn being loose. If after carrying out the above examination the trouble is not rectified the horn may need adjustment, but this should not be necessary until the horns have been in service for a long period.

Adjustment does not alter the pitch of the note, it merely takes up wear of moving parts. When adjusting the horns, short circuit the fuse otherwise it is liable to "blow". Again, if the horns do not sound on adjustment release the push instantly. When making adjustments to a horn, always disconnect the supply lead of the other horn taking care that it does not come into contact with any part of the chassis and so cause a short circuit.

Adjusting the Horn

Connect a 0–10 ammeter in series with the horn to be adjusted. The current passed by one horn only should measure approximately 4 amperes. If necessary, turn the adjusting screw (see Fig. 20) anti-clockwise to reduce the current, and vice versa. The adjusting screw should not be turned more than three notches at a time before checking the current. Repeat adjustment until the best performance is obtained with approximately the current quoted above.

WINDSCREEN WIPER

To start the electrical windscreen wiper it is only necessary to move the switch to the "On" position. Parking of the arms is effected by switching off at the end of the stroke. On no account must the arms be moved manually across the screen.

Arm and Blade Assembly

To remove the arm and blade assembly, slacken the securing nut and continue to rotate it until the extracting device embodied in the arm frees the arm from the spindle. When fitting the replacement arm and blade, slacken the securing nut and push the arm on to the spindle as far as it will go. Set the arm in its correct position and tighten the securing nut.

Replacing a Blade

To fit a new wiper blade adopt the following procedure.

Lift the arm and blade from the windscreen and press the arm and blade together. In the same movement slide the blade off the curved portion of the arm. To replace, insert the curved portion of the arm in the slot of the blade swivel and slide the blade into position.

Lubrication

Occasionally, lightly grease both sides of the curved portion of the wiper arm which fits into the slots of the blade swivel.

DIRECTION INDICATOR

In the event of trouble occurring in the flasher system the following procedure should be adopted:—

1. Check the bulbs for broken filaments.

2. Refer to the wiring diagram and check over all flasher circuit connections.

Fig. 21. This illustration shows the location on the left-hand flitchplate of the direction indicator flasher unit.

3. Switch on the ignition.

(a) Check that terminal "B" at the Flasher Unit is at 12 volts with respect to earth.

(b) Connect together terminals "B" and "L" at the Flasher Unit, and operate the direction indicator switch.

If the flasher lights now work, the Flasher Unit is defective and must be replaced.

If the flasher lights do not work, relay DB10 is defective and must be replaced.

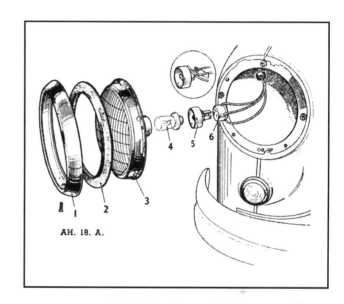

Fig. 22. Headlamp exploded.
1. Front rim. 2. Rubber seal. 3. Glass and reflector. 4. Bulb.
5. Bulb holder (U.S.A.). 6. Three pin socket. Inset shows bulb holder for all models except U.S.A.

LIGHTING

Headlamp—Replacing Bulbs

To remove the light unit for bulb replacement, un-screw the screw securing the front rim and lift off the rim. Next remove the dust-excluding rubber when three spring-loaded adjustment screws will be visible. Press the light unit in against the tension of the adjusting screw springs and turn it in an anti-clockwise direction until the heads of the screws can be disengaged through the slotted holes in the light unit rim. Do not disturb the screws as this will alter the lamp setting.

Twist the adapter in an anti-clockwise direction and pull it off. The bulb can then be removed.

Put the replacement bulb in the holder taking care to locate it correctly. Engage the projections on the inside of the adapter with the slots in the holder, press on and secure by twisting in a clockwise direction.

Position the light unit so that the heads of the adjusting screws protrude through the slotted holes in the flange, press the unit in and turn in a clockwise direction. Replace the dust-excluding rubber and refit the front rim.

Setting

The lamps should be set so that the main driving beams are straight ahead and parallel to one another, and parallel to the road surface. If adjustment to the setting is required, first remove the front rim and rubber as previously described. Set each lamp to the correct position in the vertical plane by means of the vertical adjustment screw at the top of the reflector unit. Turn the screw in a clockwise direction to raise the beam and in an anti-clockwise direction to lower it. Horizontal adjustment can be altered by turning the adjustment screws on each side of the light unit.

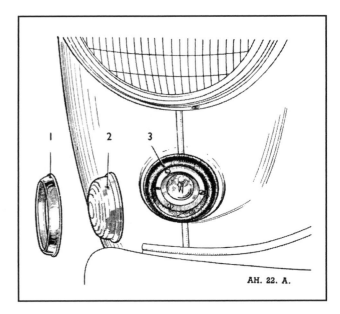

AH. 22. A.

Fig. 24.   Combined side and flasher lights.
1. Light rim.   2. Glass.   3. Rubber lip.

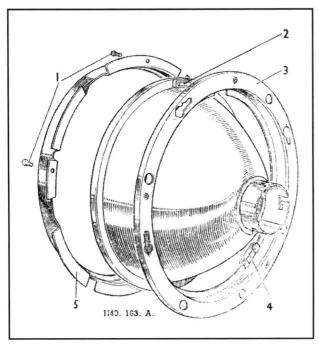

Fig. 23.   Headlamp reflector assembly.
1. Securing screws.   2. Die cast projection.   3. Seating rim.
4. Packing clip.   5. Unit rim.

The setting of the lamps can best be carried out by placing the car in front of a blank wall at the greatest possible distance, taking care that the surface on which the car is standing is level and not sloping relative to the wall.

It will be found an advantage to cover one lamp while setting the other.

Replacing a light unit

In the event of damage to either the front lens or reflector, a replacement light unit must be fitted as follows:—

1. Remove the light unit as already described.

2. Withdraw the three screws from the unit rim and remove the seating rim and unit rim from the light unit.

3. Position the replacement light unit between the unit rim and setting rim, taking care to see that the die cast projection at the edge of the light unit fits into the slot in the seating rim, and also check that the seating ring is correctly positioned. Finally secure in position by means of the three fixing screws.

Cars for Export to U.S.A.

In order to comply with the lighting regulations in certain States, a sealed beam unit must be fitted in place of the Lucas light unit.

Cars intended for the American market are fitted with special headlamp bulb adaptors and Ward and Goldstone sockets.

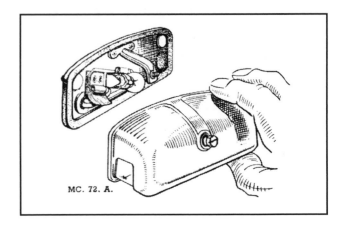

MC. 72. A.

Fig. 25.   Number plate illumination with cover removed for access to bulb.

When replacing a Lucas Light Unit by a Sealed Beam Unit, it is only necessary when connecting up to withdraw the Lucas adapter from the Ward and Goldstone socket. The socket can then be plugged directly to the Sealed Beam Unit.

Side, Stop-Tail Lamps and Flasher Lights

To gain access to the bulb move aside the rubber ring and lever off the rim from the bottom of the lamp. Move aside the inner rim and remove the glass in a similar manner.

Note that these bulbs can only be inserted in the holder in one position. This ensures that the 6-watt filaments can be used only as Side or Tail lights and the 18-watt filaments only as Stop or Direction Indicator lights.

To refit the glass, move aside the rubber rim and locate the glass at the top of the lamp. Press the glass into position and fit the rubber ring round it. The outer rim is fitted in the same way.

Number Plate Illumination

By slackening the single securing screw and removing the front cover assembly, access is gained to the bulb fitting.

LOCATION AND REMEDY OF FAULTS

IGNITION CIRCUIT

1. Engine will not Fire

(a) See that the battery terminals are secure and that the battery is in a charged condition, either by the use of a hydrometer or by checking that the starter will turn the engine and the lamps give good light.

If the battery is discharged, it must be recharged from an independent electrical supply.

(b) Ensure that the controls are correctly set for starting.

(c) Remove the cable from the centre distributor terminal, hold it so that the end is about ¼ in. away from some metal part of the chassis while the engine is turned over slowly. If sparks jump the gap regularly, the coil and distributor are functioning correctly and the sparking plugs must be examined. If these are clean and the gaps correct, the trouble is due to carburettor, fuel supply, etc.

(d) If the coil does not spark on test (c), check for a fault in the low tension wiring. This will be indicated if no spark occurs between the distributor contacts when quickly separated by the fingers when the ignition is switched on.

Examine all cables in the ignition circuit and see that all connections are tight.

(e) If the wiring proves to be in order, examine the distributor.

Spring back the clips on the distributor head and remove the moulded cover. Lift off the rotor, carefully levering with a screwdriver if necessary.

Check the contacts for cleanliness and correct gap setting as described under "Distributor Cleaning", pages O/2–O/3.

Connect a 6-watt test lamp between the distributor low tension terminal and earth. Turn the engine by engaging top gear and rolling the car forward. The lamp should flash with the opening and closing of the contacts.

If it lights but does not flash, badly adjusted or dirty contacts are indicated. If it does not light, a broken or loose connection, or contacts remaining closed, is indicated.

2. Engine Misfires

(a) Examine the distributor contacts, if necessary cleaning them and adjusting the gap, as already described.

(b) If a sparking plug is suspected of cutting-out, a quick test may be made with the engine running by lightly holding the porcelain portion of each plug in turn and noting any difference in temperature. The defective plug will be comparatively cool. If time is not at stake, the best method is to substitute a spare for each sparking plug, in turn, until the engine fires evenly.

However, if this test fails to provide the cure, the offending cylinder can be traced by momentarily removing and replacing each plug lead in turn. When the lead from the sparking plug of the defective cylinder is removed, there will be no noticeable change in the engine running note. On others, however, there will be a pronounced increase in roughness.

Examine the cable from the plug to the distributor cover for deterioration of the insulation, renewing the cable if the rubber is cracked or perished. Clean and examine the distributor moulded cover for free movement of the carbon brush. If a replacement brush is necessary see that the correct type is used; the standard non-resistive brush is too short for use with this distributor and will not make contact with the rotating electrode.

If tracking has occurred, indicated by a thin black line, usually between two or more electrodes, a replacement distributor cover must be fitted.

(c) If sparking is regular at each plug when tested, the trouble is probably due to engine defects or the carburettor, petrol supply, etc.

3. Low Tension Circuit

If it is determined that the fault lies in the low tension circuit, by the eliminating check (1e) switch on the ignition and turn the engine until the contact breaker points are fully opened.

Refer to the wiring diagram and check the circuit with a voltmeter (0–20 volts) between the following points and a *good* earth.

If the circuit is in order, the voltage reading should be approximately 12 volts. No reading indicates a damaged cable or loose connections, or a breakdown in the section under test.

Battery to Control Box: Connect the voltmeter between control box terminal "A" (brown cable) and earth. No reading indicates a faulty cable or loose connection. This section of circuit is made via the Solenoid Starter Switch. In the event of zero reading, also check voltage at this point.

Control Box: Check the voltage to earth at control box terminal "A1". No reading indicates a defective series winding.

Control Box to Lighting Switch: Connect the voltmeter between lighting switch terminal "A" (brown with blue cable) and earth. No reading indicates a faulty cable or loose connection.

Lighting Switch to Ignition Switch: Connect the voltmeter between the ignition switch terminal to which the brown with blue cable is connected and earth. No reading indicates a faulty cable or loose connection.

Ignition Switch: Check the voltage to earth at the second ignition switch terminal. No reading indicates a defective switch.

Ignition Switch to Ignition Coil: Remove the (white) cable from the ignition coil terminal "SW" and connect the voltmeter between the free end of this cable and earth. No reading indicates a faulty cable or loose connection. If a reading is obtained, remake the connection to the coil.

Ignition Coil: Remove the (white with black) cable from the ignition coil terminal "CB" and connect the voltmeter between this terminal and earth.

No reading indicates a defective primary winding. If a reading is obtained, remake the connection to the coil.

Ignition Coil to Distributor: Disconnect the low tension cable from the distributor and connect the voltmeter between the free end of this cable and earth. No reading indicates a faulty cable or loose connection. If a reading is obtained, remake the connection to the distributor.

Contact Breaker and Capacitor: Connect the voltmeter across the contact points. If no reading is obtained, recheck with the capacitor removed. If a reading is now given, the capacitor is faulty and must be replaced.

Measure the contact breaker spring tension. This should be 20–24 oz., measured at the contacts.

4. High Tension Circuit

If the low tension circuit is in order, remove the high tension lead from the centre terminal of the distributor cover. Switch on the ignition and turn the engine until the contact closes. Flick open the contact breaker lever whilst the high tension lead from the coil is held about $\frac{3}{16}$ in. from the cylinder block. If the ignition equipment is in good order, a strong spark will be obtained. If no spark occurs, a fault in the circuit of the secondary winding of the coil is indicated and the coil must be replaced. The high tension cables must be carefully examined, and replaced if the rubber insulation is cracked or perished, using 7 mm. rubber covered ignition cable, see page O/3.

The cables from the distributor to the sparking plugs must, of course, be connected in the correct firing order.

5. Contact Breaker Mechanism

Check and adjust as previously detailed on page O/3. Ensure that the moving arm is free on the pivot. If sluggish, remove the arm and polish the pivot pin with a strip of fine emery cloth, lubricate with a spot of clean engine oil and replace the arm.

CHARGING CIRCUIT

1. Battery in low state of charge

(a) This state will be shown by lack of power when starting, poor light from the lamps and hydrometer reading below 1.200, and may be due to the dynamo either not charging or giving low or intermittent output. The ignition warning light will not go out if the dynamo fails to charge, or it will flicker on and off in the event of intermittent output.

(b) Examine the charging and field circuit wiring tightening any loose connections or replacing broken cables. Pay particular attention to the battery connections.

(c) Examine the dynamo driving belt: take up any undue slackness by turning the dynamo on its mounting.

(d) If the cause of the defect is not apparent, have the equipment examined by a Lucas Service Depot or Agent.

2. Battery overcharged

This will be indicated by burnt-out bulbs, very frequent need for topping-up of the battery, and high hydrometer readings. Connect an ammeter between dynamo and battery by withdrawing both leads from the "A" terminal of the control box and wiring them both to one of the ammeter terminals. From the second ammeter terminal run a cable to the control box terminal "A".

Check the ammeter readings when the engine is running steadily (1,500 r.p.m. approx.) with a fully charged battery and no lights or accessories in use—the charge reading should be within the region of 3–4 amperes. If the ammeter reading is in excess of this value, it is advisable to have the regulator setting tested and adjusted if necessary by a Lucas Service Depot or Agent.

STARTER MOTOR

1. Starter Motor lacks power or fails to turn engine

(a) See if the engine can be turned over by hand (spanner on crankshaft pulley nut). If not, the cause of the stiffness of the engine must be located and remedied.

(b) If the engine can be turned by hand, first check that the trouble is not due to a discharged battery.

(c) Examine the connections to the battery, starter and starter switch, making sure that they are tight and that the cables connecting these units are not damaged.

(d) It is also possible that the starter pinion may have jammed in mesh with the flywheel, although this is by no means a common occurrence. To disengage the pinion, rotate the squared end of the starter shaft by means of a spanner.

2. Starter operates but does not crank engine

This fault will occur if the pinion of the starter drive is not allowed to move along the screwed sleeve into engagement with the flywheel due to dirt having collected on the screwed sleeve. Clean the sleeve carefully with paraffin.

3. Starter pinion in mesh when engine is running

Stop the engine and see if the starter pinion is jammed in mesh with the flywheel, releasing it if necessary by the rotation of the squared end of the starter shaft. If the pinion persists in sticking in mesh, have the equipment examined at a Service Depot. Serious damage may result to the starter if it is driven by the flywheel.

LIGHTING CIRCUITS

1. Lamps give insufficient illumination

(a) Test the state of charge of the battery, recharging it if necessary either by a long period of daytime running or from an independent electrical supply.

(b) Check the setting of the lamps.

(c) If the bulbs are discoloured as the result of long service, they should be replaced.

2. Lamps light when switched on, but gradually fade out
As paragraph 1 (a).

3. Brilliance varies with speed of car

(a) As paragraph 1 (a).

(b) Examine the battery connections, making sure that they are tight, and replace faulty cables.

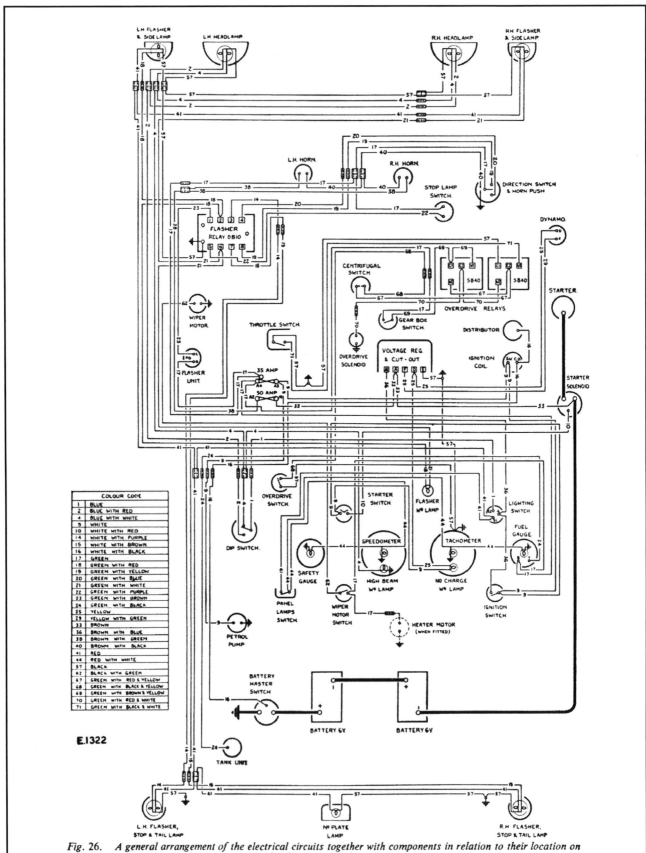

Fig. 26. A general arrangement of the electrical circuits together with components in relation to their location on the chassis and bodywork.

SERVICE DIAGNOSIS GUIDE

Symptom	No.	Possible Fault
(a) Engine will not Start		See section D, page 27
(b) Engine Misfires		See section D, page 27
(c) Battery in low state of charge, shown by lack of power when starting. (Hydrometer readings less than 1.200)	1 2 3 4 5 6 7	Fan belt tension too slack Cut-out contacts not closing Loose or broken connection in circuit Commutator greasy or dirty Brushes worn, not fitted correctly, or wrong type Regulator not functioning correctly Dirty cut-out contacts
(d) Battery overcharged, shown by burnt out bulbs and very frequent need for 'topping-up'. Hydrometer readings high	1	Generator giving high output due to regulator not functioning correctly
(e) Starting motor lacks power or fails to turn engine	1 2 3 4 5	Tight engine Discharged battery Broken or loose connection in starting motor circuit Internal fault in starting motor Starter pinion jammed in mesh with flywheel
(f) Starter pinion will not disengage from flywheel when engine is running		Check 5 in (e)
(g) Lamps give insufficient illumination	1 2	Check 2 in (e) Lamps out of alignment Bulbs discoloured through long service
(h) Lamps light when switched on, but gradually fade out	1	Check 2 in (e) Defective battery
(i) Brilliance varies with speed of vehicle	1	Check 2 in (e) Battery connections loose or broken
(j) Lights Flicker	1	Loose connections
(k) Failure of Lights	1 2	Check 2 in (e) Bulb failure Faulty cable or connections
(l) Failure of headlamps to dip	1	Defective dip switch

BODYWORK

THE careful maintenance of the bodywork, both internally and externally, is of primary importance if the car is to retain its appearance and comfort. The paintwork, upholstery, carpets, door locks, hinges, etc., will each benefit from the periodical attentions briefly indicated in the following paragraphs, and these attentions should therefore be undertaken as regularly as possible.

CARE OF BODYWORK

Cellulose

Models which are finished in cellulose should receive regular care and attention to maintain their lustre. If neglected the paintwork will become dull and develop a bloom which is the result of exhaust fumes and other traffic vapours.

Frequent washing with clear cold running water will help to maintain its lustre but ensure that the sponge and leather used are always clean and not the same mediums used on the undercarriage.

Should the paint finish become dull, a sparing application of a reputable liquid polish will restore it.

Synthetic Enamel

Cars which have a synthetic enamel finish require frequent washing to retain the high degree of finish and a good emulsion polish, occasionally applied, is sufficient to preserve the paintwork. Never use a wax liquid polish on synthetic enamel.

Chromium

Plated parts should be finished with a damp leather. If very dirty wash the chrome with warm soapy water, but on no account use metal polish.

Door Locks and Hinges

Occasionally apply a few drops of oil on the moving parts of all door locks and hinges. A light touch of grease should be smeared on the lock striker plates to ensure free movement and reduce wear of the locks.

In addition, the security of door hinges, locks, dovetails and striker plate should be checked periodically with a screwdriver.

Upholstery

The leather work has, in general, an impermeable surface and it can be kept clean and fresh looking by an occasional wiping down with a damp cloth and saddle soap. Finish off when completely dry with a good furniture cream.

Carpets are best kept in good condition by brushing or by use of a vacuum cleaner. Periodically the carpets should be removed and thoroughly beaten after which the body floor should be inspected for signs of rust.

Such spots should be cleaned and painted with a quick drying enamel before the carpets are replaced.

Windows

Windscreen, sidescreens, rear window and driving mirror should be cleaned with a damp leather.

DISMANTLING AND ASSEMBLING BODY PARTS

Fig. 1. Points 1 and 2 illustrate the nuts and bolts that are withdrawn from the hinges for bonnet removal.

Bonnet Top

The bonnet at its front edge, has two brackets which form part of the hinges. A leg from the shroud is secured to each bracket by two nuts and bolts, therefore removal of the bonnet top is achieved by withdrawing the bolts from each bracket then lifting off the panel.

Opening of the bonnet has been described in the section "Instruments and Controls".

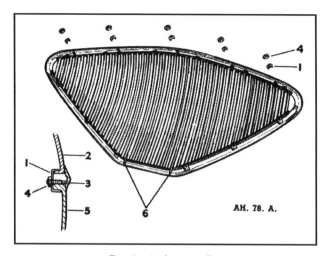

Fig. 2   Radiator grille.

1. Clamping washer.	4. Securing nut
2. Cowl section.	5. Grille section.
3. Grille stud.	6. Locating pegs.

Inset shows section of assembled grille.

Grille

Before attempting to remove a grille it is advisable to remove either one of the front road wheels. This will allow the operator to reach through the wheelarch aperture and so gain access to the five nuts that screw on to the studs of the top rail of the grille. By taking off the special washer, see Fig. 2, from each stud, the top of the grille can be carried forward clear of the cowl then lifted to withdraw the grille locating pegs that drop into holes in the lower flange surrounding the grille aperture.

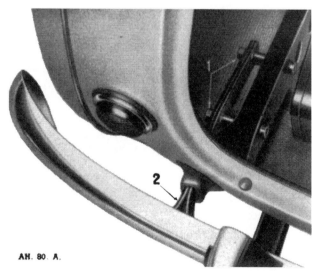

AH. 80. A.

Fig. 3.   Rear bumper fixing.

1. Fixing bolts.	2. Supporting brackets.

Bumper Bars

Rear: The rear bumper is best dismantled by releasing the two nuts immediately behind the bumper bar at the junction of bracket and bumper. The brackets

supporting the bumper are secured by two setpins to each chassis frame side member accessible within the luggage compartment.

Front: In a manner similar to that employed for its counterpart at the rear, the front bumper bar is held by two nuts to the supporting brackets which are secured by two setpins to each chassis side member.

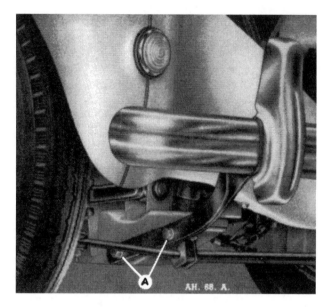

Fig. 4.   Front bumper fixing.
A. Bracket to chassis setpins.

Front Outer Apron

Once the front bumper has been removed, the outer apron can readily be released from the bodywork. There are only six cross head bolts and nuts with their spring and plain washers securing the apron to the cowl and wing assembly.

Inner Apron

Take out the radiator grille as already described. Disconnect the "flasher" unit cable at the snap connector

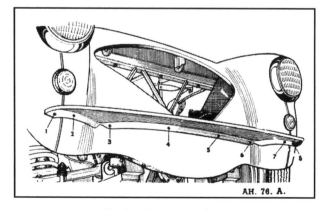

AH. 76. A.

Fig. 5.   Outer apron fixing.
Numbers 1 to 8 are the heads of the cross-head screws securing outer apron to both wings and cowl assembly.

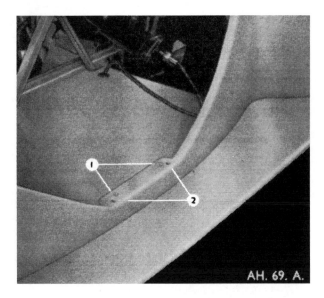

Fig. 6. *Inner apron fixing.*
1. Cross head screws.   2. Locating holes for cowl pegs.

beneath the apron and thread the cable back through the hole in the apron.

Having released the nuts and washers beneath the apron withdraw the two bolts (cross heads) that pass through the cowl lip at the grille opening. Finally, extract the setpins of each horn bracket, which will release the rear of the apron, and lift out the apron.

Front Wing

The first operation in dismantling a front mud-wing is to remove both the head and sidelight concerned; details of lens and reflector unit removal procedures are given in the "Electrical Equipment" section of this

Fig. 7. *Inner apron lower fixing.*
1. and 2. Horn and apron set pins.   3. Cowl nuts.

manual. The outer case of the headlamp is held by four bolts with brass nuts accessible beneath the wing. The sidelamps are secured by three cross head bolts and nuts.

Beneath the headlight aperture there are three bolts which secure the wing to the cowl centre. These bolts screw into spring-clip type nuts. Along the top edge of the wing flange and forward of the scuttle, four bolts, screwing into clip nuts, clamp the wing to the bonnet

surround with an additional metal thread screw, that holds an electrical harness to the surround flange, and passes into the wing flange.

In the cockpit, behind the fascia, there are a further three setpins that screw into caged nuts on the wing. These $\frac{1}{4}$ in. nuts and bolts secure the lower flange of the wing to the underside of the scuttle.

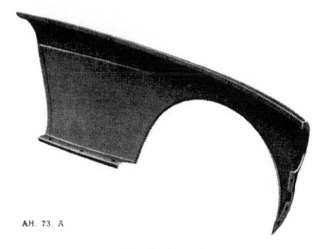

Fig. 8. *The front wing.*
Securing holes may be seen at top, bottom and front flanges.

Before the wing can be finally removed there are a number of metal thread screws to be extracted that fix the wing, on the inside of the door pillar. The door must be opened to gain access to these screws.

Rear Wing

First remove the rear wheel concerned when it will be discovered that each rear wing is fixed to the main bodywork structure by six square head bolts with spiral clip nuts which are located over the wheelarch and round the rear curve of the wing.

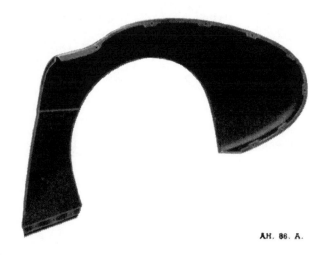

Fig. 9. *Rear wing securing flanges.*

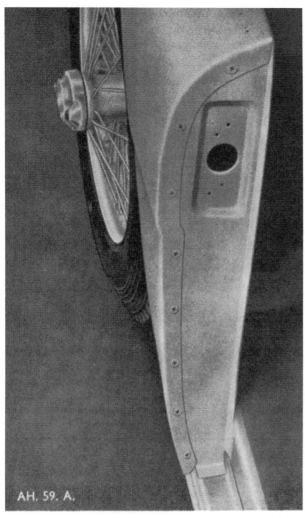

Fig. 10.   Rear wing cross-head bolts at door pillar.

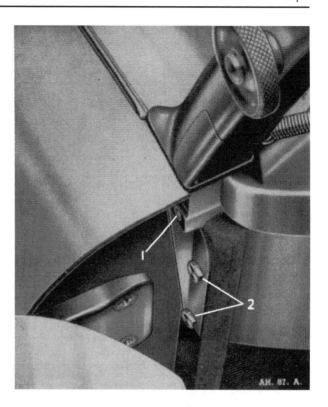

Fig. 11.   Windscreen pillar fixing.
1. Tapped hole for setpin from under fascia.
2. Nuts and bolts.

Fig. 12.   Hood frame fixing to quarter panel.

At the top of the wheelarch, head accessible within the luggage compartment, there is a plain nut and bolt, with washers, to be extracted. Within the cockpit, with its countersunk head hidden by the quarter casing, is another bolt screwing into a caged nut. As the quarter casing must be removed to gain access to this fixing point the hood frame will need to be released, the fixing bolts of which assist in securing the wing.

At the lower front edge of the wing, where its flange is secured to the chassis, there are two nuts and bolts and a vertical drive screw.

To complete the wing dismantling extract the eight $\frac{3}{16}$ in. countersunk cross-head nuts and bolts with their plain and spring washers that fix the wing leading edge to the door pillar.

Windscreen

The windscreen frame is secured to the scuttle at each side by two nuts and bolts and a single setpin. Each nut and setpin head are accessible within the cockpit behind the fascia. The bolt heads can be seen at the door pillars when the doors are open.

Hood Frame

The hood frame is secured at each side to the rear quarter panel, immediately behind the seats, by three bolts, which have cross head countersunk heads, nuts and washers. The latter are accessible beneath the wing.

With these bolts withdrawn the hood frame complete with its fabric can be removed from the bodywork.

Cockpit Moulding

The cockpit front moulding is secured to the scuttle by four cross head screws. At the rear of the cockpit

Fig. 14. The 5 drive screws that secure the shroud at the rear of the bonnet opening.

the aluminium moulding is held in place by five cross head screws.

For the door top edge moulding see section entitled "Doors".

Shroud

The shroud is not removed for normal maintenance work, however, if it should become necessary to remove the shroud due to damage the following fixing points must be made free.

Each outer wing half should be dismantled, see "Front Wing" also the front bumper, grille, inner and outer apron, windscreen and driving mirror complete with bracket, also the cockpit moulding. In addition the bonnet top must be removed thus giving access to the fixing points around the perimeter of the opening to the engine compartment.

Fig. 13. With the bonnet open 5 cross head bolts and nuts are visible which secure the shroud to the body front cross bracing.

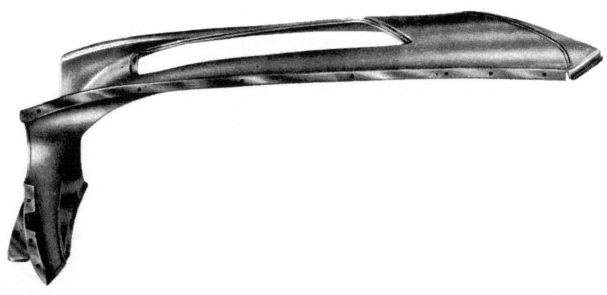

Fig. 15. The complete shroud, removed from the bodywork, showing the side fixings.

Fig. 16. Shroud upright brace with two cross-head bolts at A.

At the rear of the opening there are five drive screws holding the shroud to the scuttle. At the front of the bonnet opening five cross head bolts and nuts secure the shroud to the front cross bracing of the bodywork. Still working within the opening, at each side, there are two countersunk cross headed bolts with nuts that fix the shroud to upright braces from the chassis frame.

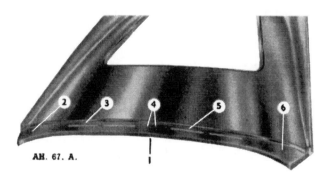

Fig. 17. Shroud rear fixing.
1. *Five pop-rivet holes on lip.*
2. *and 6. Holes for tonneau cover studs.*
3. *and 5. Demister ducts.*
4. *Fixing holes for driving mirror.*

Securing the shroud to each wheelarch panel there are two plate brackets from which the two nuts and bolts must be extracted. The cowl, which is part of the shroud, has two brackets that secure the body member to the frame dumb-irons. From these brackets extract the two nuts and bolts.

Fig. 18. Sidescreen socket and extractor tool.

Finally free the rear end of the shroud. This is secured to the scuttle just above the fascia by five "pop" rivets, with a further "pop" rivet and two soft rivets at each side fixing the shroud to the scuttle.

The complete shroud can now be lifted clear of the frame and remainder of the bodywork.

Sidescreens and Sockets

The sidescreens each have two locating dowels at their base. These dowels are a snug push fit into sockets let into the top of each door.

If necessary the sidescreen sockets can be screwed out of the door using either a broad blade screwdriver or, preferably, a special tool which incorporates a pilot, see Fig. 18.

Fig. 19. Door hinges and check strap with the link pin at A.

*Fig. 20. The door top moulding is held in place
by three cross-head screws A.*

Doors

Hinges and Door Removal: Both the upper and lower
hinge of each door is secured to the door post by four
cross head screws plus one hexagon head setpin. At the
door frame each hinge is fixed by four cross head screws.

There is a check strap fitted to each door which
must be released when dismantling a door from the
bodywork. This check strap can be released by withdrawing the small screw from the coupling near to the door
shell. Thus with the door wide open, the hinges can
readily be uncoupled from the door and the door removed.

Casing: Each door casing, complete with its trim-
ming, can be removed from the door shell after its
fourteen securing cross head screws have been withdrawn
from around the casing perimeter. One of these screws
also holds the chrome lock plate.

Door Top Moulding: The aluminium moulding at
each door top edge is held in place by three cross head
screws.

Locks: Each lock unit is fixed to the door frame by
four countersunk head bolts which have nuts on the
inside of the door shell.

The door "pull-strap" is secured to the door lock
handle at the stem end by a nut. Similarly the forward
end of the strap is secured to the door frame by a nut
and bolt; both nuts being accessible on the inside of the
door pocket.

Fascia Panel

Instrument Board: There are eight screws, the heads
of which are behind the fascia, passing through the fascia
panel into tapped holes of brackets which are an integral
part of the instrument board. By extracting these screws
(cross heads) the instrument panel can be brought through
the fascia into the cockpit thus giving access to the rear
of each instrument. On later models there is no
separate instrument board; the instruments being fitted
direct to the fascia panel.

Fascia: The main fascia panel is fixed in place by
two nuts and bolts through scuttle brackets at each
extreme end. Centrally, to the scuttle there is one metal
thread screw with a cup washer beneath its head. Below
this screw a 2 BA. nut and bolt fixes the fascia to the
choke control bracket.

By releasing these fixing points the fascia can be
relieved of the scuttle.

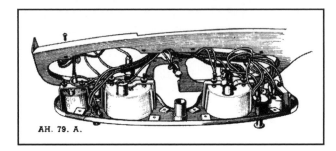

*Fig. 21. The instrument panel dismantled
from the fascia board.*

Grab Handle

The passenger grab handle is fixed to the fascia
panel by two round head screws the heads of which are
situated behind the panel.

*Fig. 22. Seat fixing of early models
showing the two setpins and alternative
holes for positioning.*

Seats

To adjust or remove either seat the cushion must be lifted whereupon the heads of two setpins are revealed. These setpins (one each side of the seat frame) secure the seat to the body floor. On their extraction the seat may be removed or repositioned, there being four alternative holes for adjustment each side of the seat frame.

On later models an adjustable driving seat is provided for forward or rearward positioning by pushing the lever, beneath the seat, toward the runner then moving the seat to the required setting and releasing the lever.

Lift out the seat cushion to gain access to the six nuts securing the seat to the runners.

The seat runners, with their packing pieces, are bolted to the floor, each runner having three bolts with nuts accessible beneath the floor.

The passenger seat of the later model is non-adjustable, but whilst being removable in the manner described for the early type car, its packing pieces are secured to the floor boards by through bolts and nuts.

Gearbox Cover

The gearbox cover, or tunnel, is secured at each side flange by four metal thread screws to the floor boards. The heads of these screws are hidden from view until each side carpet is peeled back.

Immediately before the tunnel there is a carpet covered bulkhead plate which can be removed for further access to gearbox and clutch housing. This is fixed to the bulkhead by six self-tapping screws.

Quarter Casing

Having first removed the hood frame securing bolts,

AH. 77. A.

*Fig. 23. The gearbox cover is secured
to the floor boards by seven screws.*

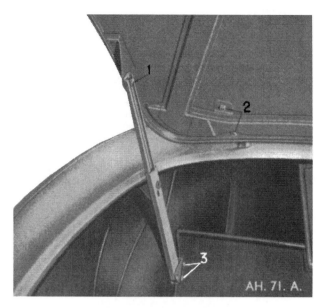

AH. 71. A.

Fig. 24. Boot lid prop rod.
1. Self locking nut.   2. Hinge nuts.
3. Prop rod bracket nuts.

withdraw the three cross head self-tapping screws, also the single stud fastener screw, then the quarter panel casing can be removed from the side concerned.

Boot Lid

To free the boot lid from its fixings, it is only necessary to undo the single self-locking nut at the top of the lid "prop-rod" also the single nut and washer at each hinge beneath the surrounding panel.

For further dismantling each hinge may be taken off the boot lid by releasing the two hinge nuts from the underside of the lid.

The locking handle can be withdrawn through lock and compartment lid after its two securing screws are extracted. The round heads of these screws are visible on the underside of the lid.

Four cross headed screws with nuts and washers fix the lock assembly to a supporting bracket riveted and welded to the lid. There is little that can go wrong with this lock, however it does benefit from the occasional application of oil particularly to the spring that is partially visible at the closing edge of the boot lid.

The lid "prop-rod" can be taken from the left-hand side of the luggage compartment by removing the two nuts and spring washers fixing the rod to its anchoring bracket.

Rear Body Panel

This panel, which forms the lower rear part of the luggage compartment, is the most likely panel to suffer damage in the event of a rear collision.

Should the rear panel require replacing, the damaged part can be removed with a minimum amount of cutting, but a number of rivets have to be drilled out.

At each side the panel is fixed by two "pop-rivets" and by two of the mud-wing nuts and bolts. The top edge of the panel has thirteen rivets holding it to the luggage compartment frame and the lower lip has nineteen rivets for securing.

After the rivets and nuts and bolts have been freed, the rear light should be removed allowing the panel to be cut along the welded seam at each side which can be felt inside the compartment.

Naturally, the new panel must be re-welded along

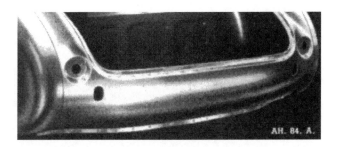

Fig. 25.   Rear body panel.
The rivets at top and bottom lips are clearly shown.

the same lines and secured by the requisite number of rivets and nuts and bolts.

PART NAME ALTERNATIVES

	Austin Part Name	Alternatives
ENGINE	Gudgeon Pin … …	Piston Pin. Small End Pin. Wrist Pin
	Scraper Ring … …	Oil Control Ring
	Welch Plug … …	Expansion Plug. Core Plug. Sealing Disc
	Oil Sump … …	Oil Pan. Oil Reservoir
CONTROLS …	Choke … … …	Strangler. Easy Starting Device
GEARBOX …	Gear Lever … …	Shift Lever
	Change Speed Fork. …	Shift Fork. Selector Fork
	First Motion Shaft …	Clutch Shaft. First Reduction Pinion. Main Drive Pinion
	Layshaft … …	Counter Shaft
AXLE … …	Crown Wheel … …	Ring Gear. Spiral Drive Gear
	Bevel Pinion … …	Small Pinion. Spiral Drive Pinion
	Spring Clips … …	"U" Bolts
	Axle Shaft … …	Half Shaft. Hub Driving Shaft. Jack Driving Shaft
STEERING …	Swivel Pin … …	Pivot Pin. Steering Pin. King Pin
	Swivel Axle … …	Stub Axle
	Cross Tube … …	Tie Rod. Track Rod
	Side Tube … …	Drag Link. Steering Connecting Rod
	Steering Arm … …	Drop Arm
ELECTRICAL …	Dynamo … …	Generator
	Voltage Regulator …	Control Board. Cut Out. Voltage Controller.
EXHAUST …	Silencer … … …	Muffler
BODY … …	Bonnet … … …	Hood
	Mudguard … …	Fender

SERVICE FACILITIES

THE following are the official addresses of the Austin Motor Company Limited and their Subsidiary Companies overseas, to whom all Service correspondence in those areas should be addressed. In all instances the enquirer is asked, first of all, to contact his nearest appointed Austin Distributor or Dealer before writing to one of the following addresses :

England

THE AUSTIN MOTOR COMPANY LTD
Service Department,
Longbridge,
Birmingham, 31

Telephone: PRIORY 2101

Telegrams: SPEEDILY, NORTHFIELD
Cables: SPEEDILY, BIRMINGHAM

London

THE AUSTIN MOTOR COMPANY LTD
Holland Park Hall,
Holland Park,
London, W.11

Telephone: PARK 8001

Telegrams: AUSTINSERV, NOTTARCH

U.S.A.

THE AUSTIN MOTOR COMPANY LTD (ENGLAND)
Central Parts Division,
2227–9 Webster Avenue,
Bronx, 57, New York, N.Y.

Telephone: CYpress 8-4500

Telegrams: AUSTINMOTO, NEW YORK

Canada

THE AUSTIN MOTOR COMPANY (CANADA) LTD
Service Division,
Kenilworth Avenue North,
Hamilton, Ontario

Telephone: HAMILTON 4-2816

Telegrams: AUSTINETTE, HAMILTON

Australia

THE AUSTIN MOTOR COMPANY (AUSTRALIA) LTD
Joynton Avenue,
Zetland,
Sydney, N.S.W.

Telephone: FF 0321

Telegrams: AUSTINETTE SYDNEY

The Austin Motor Co. Ltd. wish to make acknowledgement to :—
Messrs. A. C. Sphinx Sparking Plug Co., Ltd.; Armstrong Patents Co., Ltd.; Borg & Beck Co., Ltd.; Champion Sparking Plug Co., Ltd.; Dunlop Rubber Co., Ltd.; Girling Ltd.; Hardy Spicer & Co., Ltd.; Joseph Lucas & Co., Ltd.; Smith's Motor Accessories Ltd.; Zenith Carburetter Co., Ltd.; for their assistance in furnishing information and illustrations for this manual where required.

LUBRICATION CHART

		REGULAR ATTENTIONS
Oil.	A	**DAILY.** Engine oil reservoir, check level
Oil.	B	**1,000 MILES (1,600 KM.)** Gearbox and Overdrive, top-up if necessary; Rear Axle, top-up if required; Steering Box, top-up; Air Cleaner, re-oil
Oil Gun.	C	Propeller Shaft bearings and Splines (3); Swivel Axles (4). Lower suspension joints (2); Steering connections (6); Steering idler. 1 filler plug; Shackle pins.' Rear end of rear road springs (2); Clutch and Brake pedals pivot lever (1); Brakes balance lever (1), rear flexible cable (1); Water Pump
Oil Can.	D	Steering column felt bush; Handbrake, brake and clutch pedal linkages; Carburetter control joints; Carburetters, replenish damper assembly oil reservoir; Distributor automatic advance, cam end drive shaft bearings, cam
Examine.	E	Brake fluid supply tank, inspect and refill to correct level; Shock absorbers, check for leaks
Oil.	F	**3,000 MILES (4,800 KM.)** Engine, drain and refill oil reservoir
Oil.	G	**6,000 MILES (9,600, KM.)** Gearbox and Overdrive, drain and refill; Rear axle, drain and refill
Grease.	H	Front hub; Dynamo bearings with H.M.P. grease
Examine.	J	Shock absorbers, top-up
Oil.	K	**9,000-12,000 MILES (14,400-19,200 KM.)** Engine Filter, renew element
Oil Gun.	L	Clutch operating shaft (2)
Grease.		Speedometer and Tachometer drives

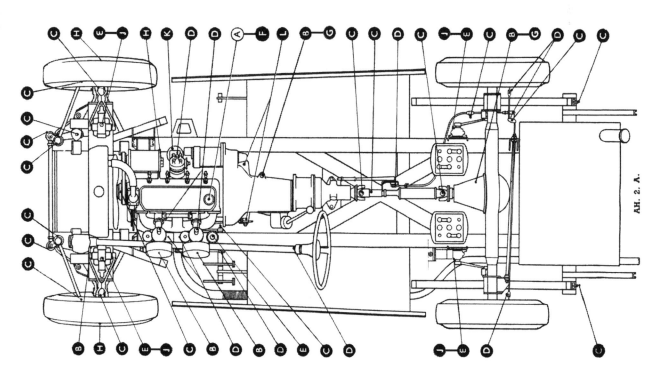

AH. 2. A.

RECOMMENDED LUBRICANTS

		Wakefield	Esso	B.P.	Duckham's	Vacuum	Shell
Engine	Summer *90° F. Down to 32° F. (Home) (32° C. to 0° C.)	Castrol XL	Essolube 30	Energol Motor Oil S.A.E. 30	Duckham's N.O.L. "THIRTY"	Mobiloil A	Shell X.100 30
Engine	Winter 32° F. +10° F. (Home) (0° C. to —12° C.)	Castrolite	Essolube 20	Energol Motor Oil S.A.E. 20 W	Duckham's N.O.L. "TWENTY"	Mobiloil Arctic	Shell X.100 20
Engine	Below 10° F. (—12° C.)	Castrol Z	Essolube 10	Energol Motor Oil S.A.E. 10 W	Duckham's N.O.L. "TEN"	Mobiloil 10 W	Shell X.100 10 W
†Transmission		Castrol XL	Essolube 30	Energol Motor Oil S.A.E. 30	Duckham's N.O.L. "THIRTY"	Mobiloil A	Shell X.100 30
Rear Axle	Down to 32° F. (0° C.)	Castrol Hi-Press	Esso XP Compound S.A.E. 140	Energol Transmission Oil E.P.–S.A.E. 140	Duckham's N.O.L. E.P.T. 140	Mobilube GX. 140	Shell Spirax 140 E.P.
Rear Axle	†32° F. to 10° F. (0° C. to —12° C.)	Castrol Hypoy	Esso XP Compound S.A.E. 90	Energol Transmission Oil E.P.–S.A.E. 90	Duckham's N.O.L. E.P.T. 90	Mobilube GX. 90	Shell Spirax 90 E.P.
†Steering Box and Oil Nipples ‡		Castrol Hi-Press	Esso XP Compound S.A.E. 140	Energol Transmission Oil E.P.–S.A.E. 140	Duckham's N.O.L. E.P.T. 140	Mobilube GX. 140	Shell Spirax 140 E.P.
Front Wheel Hubs		Castrolease Heavy	Esso Bearing Grease	Energrease C.3	Duckham's H.B.B. Grease	Mobil Hub Grease	Shell Retinax A
Distributor and Oil Can		Wakefield Oilit	Esso Handy Oil	Energol Motor Oil S.A.E. 20 W	Duckham's N.O.L. "TWENTY"	Mobil Handy Oil	Shell X.100 20
Upper Cylinder Lubrication		Wakefield Castrollo	Esso Upper Motor Lubricant	Energol U.C.L.	Duckham's Adcoids	Mobil Upperlube	Shell Donax U

* Engine: Above 90° F. or for high-speed driving at high temperatures use next heavier grade of oil.

† Transmission: For prevailing Sub-Zero (° F.) temperatures use S.A.E. 20 Lubricant.

† Rear Axle: For prevailing Sub-Zero (° F.) temperatures use S.A.E. 80 E.P. Lubricant.

‡ Steering: For prevailing Sub-Zero ° F.) temperatures use S.A.E. 80 E.P. Lubricant.

‡ Oil Nipples: For high temperature climates the grease as shown for hubs can be used.

Hydraulic Brakes: Use Girling Brake Fluid (Crimson).

Shock Absorbers: Use Armstrong's Super (Thin) Shock Absorber Oil.

Note.—Where Girling Brake Fluid and Armstrong's Shock Absorber Oils are not available the alternative used must be of the highest quality.

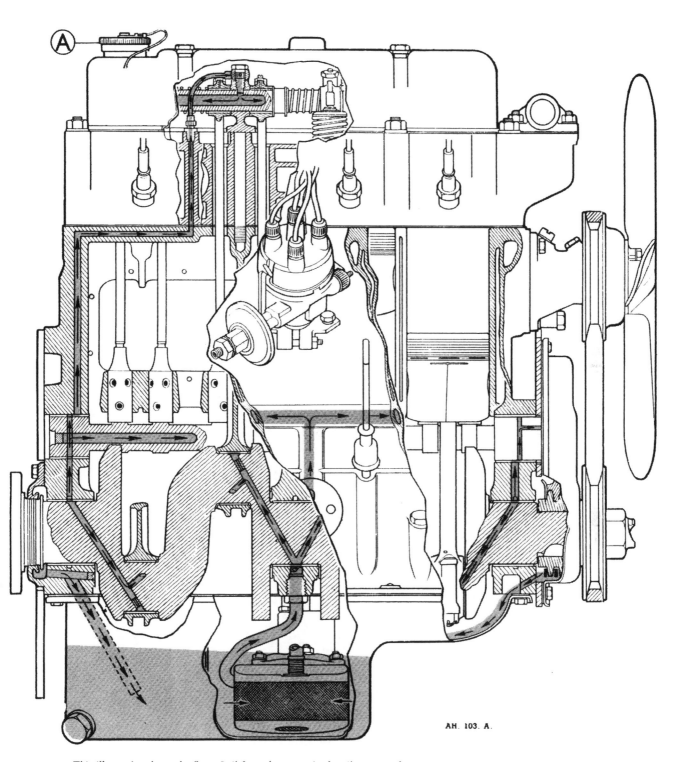

AH. 103. A.

*This illustration shows the flow of oil from the sump via the oil pump to the
main gallery, bearings and overhead rocker gear*

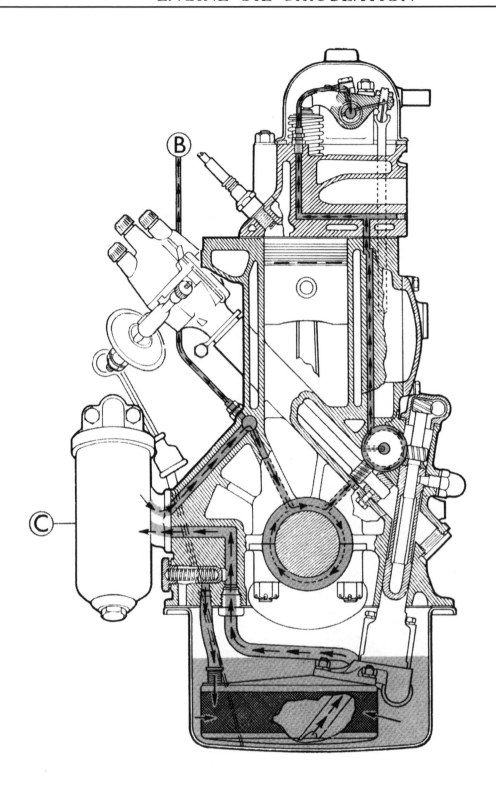

AUSTIN - HEALEY
100

SERIES . BN2

Supplement

MARCH 1956

THE AUSTIN MOTOR COMPANY LIMITED, BIRMINGHAM, ENGLAND

Introduction

THIS supplement has been written to assist owners of Austin-Healey cars, series BN2.

Many components are similar to those on the earlier models and only where major differences occur between the old and the new are they described in detail.

Where components of the old and the new models are common the operator should refer to the appropriate section in the main Austin-Healey manual, which covers the series BN1.

GENERAL INDEX

REGULAR ATTENTIONS

THE following is a convenient list of regular attentions which the car should receive to keep it in good mechanical condition. These instructions should be closely followed whether the attentions are performed by the owner or the local dealer.

The attentions under the daily or periodic headings are based on the assumption that the maximum mileage per week does not exceed 500, but see "Post Delivery Check" for special attentions during the first 1,000 miles.

Under more arduous conditions such as very dusty or muddy roads, long distances at high speeds, it will be advisable to attend to chassis lubrication more frequently.

POST DELIVERY CHECK
First 1,000 miles (1500 km.)

Austin dealers are under agreement to carry out a "Post Delivery Check" once during the period of the first 1,000 miles (1500 km.) running, or as soon as possible afterwards, of Austin vehicles purchased from them, when they will, without charge, except for materials used :—

Change the oil in the engine, gearbox and rear axle, and check the oil level in the steering box.

Lubricate all chassis points.

Check the tightness of cylinder head and manifold nuts. Tighten the fan belt if necessary.

Check tappet clearances and ignition timing.

Clean out the carburetter float chamber and check the slow running adjustment. Also clean petrol pump gauze.

Examine and adjust if necessary, the sparking plug and distributor points and verify the working of the automatic ignition control.

Check the front wheel alignment and steering connections.

Check the clutch pedal clearance.

Examine and adjust the braking system.

Check the tightness of nuts and bolts, body and bonnet cowl to chassis, spring clips, etc.

Lubricate the door locks, bolts, hinge pins and seat runners.

Test the lights, check the charging rate, wiring and terminals.

Examine the battery and bring up to the proper level with distilled water or diluted acid.

Test the tyres for correct pressure.

Road test the car after adjustments.

EVERY DAY

Engine
Check the level of oil in the sump and top up if necessary to the "full" mark on the dipstick. The oil filler is at the rear of the valve rocker cover and the dipstick is on the right-hand side of the engine.

Radiator
Check the level of the water in the radiator and top up if necessary. Fill to just below the filler plug threads when the engine is cold.

Fuel Tank
Check the quantity of fuel in the tank and add upper cylinder lubricant if desired.

Tyres
Check the tyre pressures, when the tyres are cold, not when they have reached their running temperature.

EVERY 1,000 MILES
(1500 km.)

Gearbox and Overdrive
Check the level and top up if necessary. For access take out the inspection panel in the top right-hand side of the gearbox cover when the filler plug will be accessible.

Remove the combined dipstick and filler plug, and fill up to the correct level.

Rear Axle
Check the oil level and replenish if necessary. The correct grade of oil should be injected into the axle casing from underneath the car, using the adaptor on the oil gun.

First remove the plug, which is on the rear side of the axle, then place the end of the adaptor into the oil hole, and inject with oil.

The plug also serves as an oil level indicator. Therefore do not replace the plug at once, but give the oil time to run out if too much has been injected. This is most important because if the rear axle is overfilled the lubricant may leak through to the brakes and render them ineffective. Wipe away excess oil from the casing.

Propeller Shaft Splines
Oil the nipple on the sliding yoke at the gearbox end of the propeller shaft. To get at this yoke, prise out the rubber plug situated on the rear left-hand side of the gearbox cover, after lifting up the carpet.

Universal Joints

Lubricate the propeller shaft universal joints. The front joint is best lubricated from above after the rubber plug in the gearbox cover has been removed. The rear joint may be lubricated from below or above through the hinged panel behind the seats. Move the car to bring the nipples to the required positions.

Also test the flange bolts and tighten if these have worked loose; the nuts are secured with tabbed washers on the front flange and are self-locking at the rear.

Steering Connections

Apply the oil gun to the steering centre cross tube nipples (2) and the steering side cross tube nipples (4) and top up the steering idler via the oil plug orifice.

N.B.—On no account should the steering idler be overlooked, as lack of lubricant in this component may cause a serious breakdown due to the additional load imposed on the steering box.

Steering Box

The steering box should be topped up with oil, using the special adaptor and the oil gun. Take out the hexagon plug on the side or top of the steering box to inject the oil. Make certain that grit does not enter the casing during the operation and wipe away any excess oil. To facilitate this attention it is advisable to remove the road wheel.

Steering column

Lubricate the felt bush at the top of the steering column with a few drops of light machine oil. To gain access to the bush there is a lubrication hole in the steering wheel hub.

Swivel Axles

Apply the oil gun to the two nipples on each swivel axle. This is best done when the vehicle is partly jacked up, as the oil is then able to penetrate to the thrust side of the bearings.

Shackle Pins

These are on the rear end of the rear road springs and should be given a charge of oil with the oil gun. There are two nipples, one on each bottom shackle.

Clutch Pedal

With the oil gun, lubricate the nipple at the pivot of the lever.

Brakes

Check the brakes and adjust if necessary. Apply the oil gun to the balance lever on the rear axle, the handbrake pivot, the lubricator on the flexible cable and the pedal pivot nipple.

Brake Fluid Supply Tank

Inspect and refill to the correct level, which is an inch from the top of the container. See Lubrication Chart for recommended fluid.

Brakes and Controls

With the oil can, lubricate all the handbrake linkage points, brake and clutch pedal linkages and carburetter control joints. See Lubrication Chart page K2.

Shock Absorbers

Ensure that there are no visible signs of leakage and that the rubber bushes are undamaged.

Carburetters

Remove suction chamber cap and damper assembly and replenish oil reservoir as necessary.

Battery

Ascertain the state of charge of the two 6 volt batteries by taking hydrometer readings. The specific gravity readings should be :

Fully charged ...	...	1·280 to 1·300
Half charged ...	...	approx. 1·210
Discharged ...	...	below 1·150

These figures are for an assumed electrolyte temperature of 60°F.

Check that the electrolyte in the cells is level with the tops of the separators. If necessary add a few drops of distilled water. Never use tap water as it contains impurities detrimental to the battery.

Never leave the batteries in a discharged condition. If the vehicle is to be out of use for any length of time, have the batteries removed and charged about once a fortnight.

Wheels and Tyres

Tighten the hub caps and check the tyre pressures, including the spare, using a tyre gauge. Inflate if necessary and see that all valves are fitted with valve caps. Inspect the tyres for injury and remove any flints or nails from the treads. Ensure that there is no oil or grease on the tyres since these substances are harmful to rubber.

EVERY 3,000 MILES
(5000 km.)

Engine

Drain the sump and refill with new oil. Capacity is 11¾ pints (8·71 litres). The oil filter must also be drained and then removed from the engine for refilling. Capacity is 1¼ pints (0·71 litres).

Bonnet Lock

Lubricate with the oil can the bonnet catch and safety device, and adjust if necessary.

Fan Belt

The fan belt must be sufficiently tight to prevent slip at the dynamo and water pump, yet it should be possible to move it laterally about half an inch each way.

To make any necessary adjustment, slacken the bolts and raise or lower the dynamo until the desired tension of the belt is obtained. Then securely lock the dynamo in position again.

Fuel Pump

The gauze filter of the fuel pump must be removed and cleaned. The filter plug is situated on the under-side of the pump head and must be unscrewed when the gauze strainer and washer will come away with the plug.

Swill the gauze in petrol and lightly blow through with dry air. On replacing the plug and filter ensure that the washer is in good condition.

Distributor Automatic Advance

Remove the distributor cap and add a few drops of engine oil through the hole in the contact breaker base through which the cam passes.

Distributor Cam-end Drive Shaft Bearings

Lubricate the distributor camshaft bearings by withdrawing the rotor arm from the top of the distributor spindle and carefully adding a few drops of thin machine oil round the screw exposed to view. Take care to refit the rotor arm correctly by pushing it on to the shaft and turning until the key is properly located.

Distributor Cam

Apply a trace of engine oil to the distributor cam. Be careful not to let any oil or dirt reach the contact breaker points.

Contact Breaker Points

Clean the contact breaker points. Cleaning of the contacts is made easier if the contact breaker lever carrying the moving contact is removed. To do this, slacken the nut on the terminal post and lift off the spring, which is slotted to facilitate removal. Before replacing smear the pivot on which the contact breaker works with clean engine oil.

Check the contact breaker setting, reset if necessary. The correct gap is ·014 to ·016 in.

Sparking Plugs

Remove the plugs and clean off all carbon deposit from the electrodes, insulators and plug threads with a stiff brush dipped in paraffin. Alternatively the plugs may be taken to the local Austin dealer for cleaning by specialised equipment.

Clean and dress the plug points and reset to the correct gap of ·025 in.

Before replacing the plugs check that the copper washers are in a sound condition. Never over-tighten a plug but ensure that a good joint is made between the plug body, the copper washer, and the cylinder head.

Renewal of sparking plugs is left to the owner's discretion as their efficient working life is variable. Use Champion N.A.8 long reach plugs.

Wheels

Change over the wheels diagonally (including the spare wheel) in order to obtain maximum service with even wear from each tyre.

EVERY 6,000 MILES
(10000 km.)

Gearbox

Drain when the oil is warm, after a run, and refill to the level of the filler plug with new oil. Capacity, 5¼ pints (2·98 litres). Also see "Overdrive".

Overdrive

This oil change is carried out in conjunction with the gearbox attention as the unit has no separate filler plug. However, when draining the gearbox, the Overdrive drain plug, must also be withdrawn. Remove Overdrive oil pump filter and clean the gauze by washing in petrol. The filter is accessible through the drain plug hole and is secured by a central set bolt.

Rear Axle

Drain when the oil is warm, after a run, and refill to the level of the filler plug with new oil. Capacity, 3 pints (1·704 litres).

Front Road Wheel Hubs

Unscrew the knock-on hub cap and, using the extractor provided, withdraw the grease cap from within the hub. Replace with grease if necessary, but do not over-lubricate as excess grease may penetrate to the brake linings.

The splines and the cone faces of both the hub and wheels, also the threads of the hub cap, should be smeared with grease.

Rear Road Wheel Hubs

These are packed with grease upon assembly and do not require greasing attention other than for wheels, splines and cones as detailed for front wheels and hubs.

Carburetters

The flow of fuel at each carburetter inlet to each float chamber should be checked and if necessary the filters in the unions should be cleaned.

To remove a float chamber disconnect the fuel supply pipe, slacken the lid retaining nut and uncouple the steel strut which connects to the induction pipe. Unscrew the chamber holding-up bolt, being careful to note the positions of the brass and fibre washers, and then take off the lid. Do not lose the float needle.

In addition, clean out each suction chamber assembly by removing the three securing screws and lifting off the body in the same plane to avoid damage to the needle.

Lift out the hydraulic damper and wash the assembly in petrol. Dry thoroughly, refit and replenish the damper with oil. When fully re-assembled lift the piston to its fullest extent, thus expelling surplus oil which lubricates the piston rod and eventually finds its way into the induction pipe.

Valves

Check valve clearances, and adjust if necessary.

Water Pump

There is a plug on the water pump housing which should be removed and a small charge of oil injected. It is better to under-lubricate than to overdo the attention.

Dynamo Bearings

Inject a few drops of oil through the hole in the end caps.

Air Cleaners

Every 6,000 miles the air cleaners should be removed, cleaned and re-oiled. To withdraw the element, first extract the central setpin with its shake-proof washer from the top plate, lift off the plate and pull out the element. Swill the element in petrol, drain, immerse in engine oil and again drain before refitting.

General Check

Examine and, if necessary, tighten all bolts and nuts such as road spring clips, shock absorber retaining bolts, and body panel bolts.

Examine other parts such as steering connections, brake rods, tubing, fuel pipe unions, etc., neglect of any of these points may be followed by an expensive repair and inability to use the car for a lengthy period.

EVERY 12,000 MILES
(20000 km.)

Clutch Operating Shaft

Lubricate the two nipples sparingly, as any excess lubricant may find its way into the clutch.

Speedometer and Tachometer Drives

Disconnect the cables at their instrument end and pull the inner members out of the casings. These should be lubricated sparingly by smearing with light grease. It is important that each drive is *not* over lubricated, otherwise damage will be caused to the instrument should the lubricant find its way into the head.

To reassemble, thread the cable concerned with a twisting movement into the casing, since this will help the cable to engage easily with its union at the gearbox or engine adapter. When this engagement is felt the cable can be pushed home so that the square end stands out approximately ⅜ in. from the casing. Connect up to instrument head.

Cooling System

Normally the cooling system should be flushed out twice annually upon the addition and removal of anti-freeze. In countries where anti-freeze is not required, however, the cooling system should be flushed out every 12,000 miles (20000 km.).

SERVICE ATTENTIONS

The following additional inspections and adjustments should be carried out periodically by an Austin dealer at the mileages mentioned. These attentions are not usually carried out by the owner driver and the tools, as supplied in the tool kit, are not sufficient for the work entailed.

EVERY 3,000 MILES
(5000 km.)

Clutch Pedal Adjustment

Check and adjust if necessary. The pedal should be depressed 1⅛ in. befoer the clutch springs are felt to be under compression.

EVERY 6,000 MILES
(10000 km.)

Shock Absorbers

Check the fluid levels and top up if necessary. The correct level is just below the filler plug threads. See page K/2 for recommended fluid. Carefully clear away all road dirt and grit from the vicinity of the filler plugs before removal.

N.B.—Where the recommended fluid is not available the following are acceptable alternatives: Shell X-100 20/20W, Wakefield's Castrolite, Mobiloil Arctic, Esso Hydraulic (Medium), Duckham's N.P.20, B.P. Energol S.A.E.20W.

Track Alignment

Check front wheel alignment, $\frac{1}{16}$ to $\frac{1}{8}$ in. toe-in taken along a horizontal line at centre height using the wheel rims as data points.

Ignition Timing

Check setting and adjust if necessary.

EVERY 12,000 MILES
(20000 km.)

Steering Box

Check for wear. This can be felt if the front wheels can be moved without creating any movement at the steering wheel.

Front and Rear Hub Bearings

Check for signs of wear.

Starter and Dynamo Commutators

Clean, also check freedom of brushes in holder.

External Oil Filter

Undo the "Full Flow" filter central fixing bolt and empty the container. Take out the old element, and replace with a new one. Tecalemit type FG.2313 or Purolator type M.F.26A, whichever is applicable.

Capacity of the filter is approximately $1\frac{1}{4}$ pints; prime before refitting.

TOP OVERHAUL

Decarbonising, Valve Grinding and Adjustment

This attention may not be needed so frequently on cars used for long journeys. As a general guide, a falling off in engine power indicates when decarbonising is due. The owner is advised to take his car to the local Austin Dealer for examination. The correct valve clearance is 012 in. with the engine hot or cold.

INSTRUMENTS AND CONTROLS

Fig. 1. Instrument Panel.

1. *Ignition switch.*
2. *Oil pressure gauge.*
3. *Starter.*
4. *Speedometer.*
5. *Direction warning light.*
6. *Tachometer.*
7. *Lighting switch.*
8. *Fuel gauge.*
9. *Overdrive switch.*
10. *Screen wiper control.*
11. *Water temperature gauge.*

AH. 14. B.

INSTRUMENTS

Speedometer

Registers the car speed and total mileage. The trip figures at the top of the speedometer can be set to zero by pushing up the knob at the bottom of the speedometer and turning it to the right.

Tachometer

This instrument indicates the revolutions per minute of the engine and thus assists the driver in determining the most effective engine speed range for maximum performance in any gear.

Oil Pressure Gauge

Indicates the oil pressure in the engine. It does not show the quantity of oil in the sump.

Ignition Warning Light

Glows red when the ignition is switched "on" and fades out when the dynamo is charging the battery.

Headlight Beam Warning Light

A red light appears when both headlights are switched on, with the two beams full ahead. The light goes out when the headlights are dipped.

Direction Flasher Warning Light

A green light starts to blink as soon as the direction indicator switch is operated. It is fitted to warn the driver that the indicator lights have been left flashing, as they cannot be seen from the driver's seat.

Fuel Gauge

Indicates the contents of the petrol tank when the ignition is switched on. When the tank is being filled switch off to stop the engine, switch on again and the needle will record the amount of fuel entering the tank.

Water Temperature Gauge

This records the temperature of cooling water circulating in the cylinder block and radiator. The correct running temperature under normal conditions should be 185° to 194°F. approximately.

FOOT CONTROLS

Accelerator

"Organ" type pedal at right-hand side.

Brakes

The centre pedal operates the hydraulic brakes on all four wheels.

Clutch

The left pedal. Do not rest the foot on this pedal when driving and do not hold the clutch out to "free wheel".

Dip Switch

If the headlights are on full, a touch on the foot dip switch alters the lights to the "dipped" position where they remain until another touch returns them to "full-on".

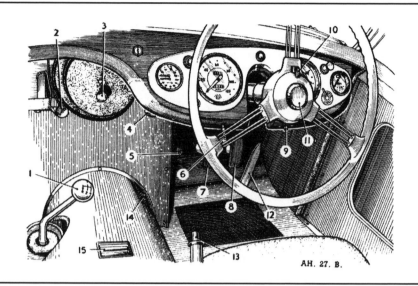

Fig. 2. Driving Controls.

1. Gear lever.
2. Choke control.
3. Heater switch.
4. Fresh air control.
5. Dip switch.
6. Speedometer trip control.
7. Clutch pedal.
8. Brake pedal.
9. Panel light switch.
10. Direction indicator switch
11. Horn button.
12. Accelerator.
13. Handbrake.
14. Gearbox filler panel.
15. Ashtray.

AH. 27. B.

HAND CONTROLS

Handbrake

Situated on the right-hand side of the propeller shaft tunnel, between the seats. To release, pull rearwards slightly and depress button, then push forward fully. Operates on the rear wheels only.

Gear Lever

Should always be left in neutral when starting the engine. The lever is mounted on the left side of the gearbox. To engage a gear, depress the clutch and move the lever to the required position.

Choke control

Located on the left side of the heater, underneath the fascia and is used when the engine is cold. Pull out to the limit until the engine fires and return it to the half-way position for rapid warming up. The choke must be released at the earliest moment possible.

Ignition Switch

Turn the key clockwise to switch on. Do not leave the switch "on" when the engine is stationary; the warning light is a reminder. The ignition key may also be used for locking the luggage compartment.

Lighting Switch

Mounted above the overdrive switch. Pull out to put on sidelights; twist to right and pull again to put on headlights. The headlights are dipped by foot operation.

Starter Switch

Press in the control to start and release as soon as the engine fires. If the engine fails to start after a few

revolutions, do not operate the starter again until the engine is stationary.

Direction Flasher Switch

The indicator lights are controlled from the centre of the steering wheel. Normally, after the car has turned a corner, the control is automatically returned to the vertical position and the lights switched off, but when only a slight turn has been made it may be necessary to return the switch manually.

Windscreen Wipers

To start the electric wipers, gently pull out the wiper control. To park, switch off by pressing the control inwards when the arms are at the end of their stroke. Do not try to push the arms across the screen by hand.

Panel Light Switch

Slide the switch to the right to illuminate the instruments. Only operates when the side lights are "on".

Overdrive Switch

A chromium plated switch is mounted on the fascia panel with two positions clearly marked. When the switch is moved to the overdrive position, the overdrive will automatically come into operation.

Heater Switch

Mounted on the heater unit which is situated below the centre of the fascia. To start, turn the switch to the right until a click is heard. The further the control is turned the less will be the speed of the fan, due to the fact that a rheostat is incorporated in the switch.

Battery Master Switch

This switch, situated in the luggage compartment recess, adjacent to the spare wheel, is fitted as an anti-thief device. The luggage compartment must of course be locked after the switch has been turned to the "off" position.

Air Intake Control

A supply of cold air, entirely independent of the heater unit, can be admitted to the car interior for ventilating purposes by pulling out the control on the right-hand side of the heater and removing the rubber grommet on the underside of the scuttle valance (right-hand side).

Horn Button

Mounted at the centre of the steering wheel operated independently of the ignition switch.

Windscreen

To gain the high speed position (see Fig. 3), unscrew the knurled knob of each pillar and push them inwards. Lift the screen to free the locating pegs and to clear the windscreen wipers, then move the windscreen forward so that the peg on each pillar fits into its forward locating hole. Finally, tighten the knurled knobs.

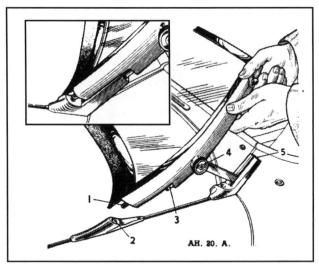

Fig. 3. Adjusting the windscreen to the high-speed position.
1. Locating peg. 2. Position for 1. 3. Locating peg.
4. Knurled locknut. 5. Location for 3.
Insert shows upright position for screen.

Hood

To stow away the hood, first release the securing locks at each windscreen pillar. The hood will then spring away from the screen. Next undo the turn buttons and stud fasteners at the rear and side of the hood, and inside the door rear pillar, then pull the steel bar of the hood rear panel out of the two chrome clips on the boot top panel.

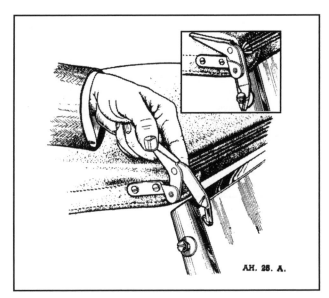

Fig. 4. An illustration showing the operation of releasing the hood securing hooks at the windscreen.
The inset shows the hook in the locked position clamping the hood rail to the windscreen.

Bring the two main ribs of the hood frame together, folding the material inwards between them. Close the trellis frames at each side of the hood so that the front rail moves rearwards to meet the first rib. Again fold the material neatly between rail and rib. With the collapsed hood upright ensure that the rear panel hangs straight down, then swing the whole assembly forward and downward to tuck away behind the seats.

Raising the hood is an exact reversal of this procedure.

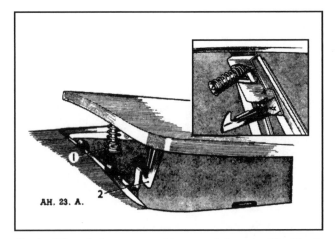

Fig. 5. When the bonnet control is operated the spring (1) will raise the bonnet until it is caught in position by the safety catch (2). Insert shows catch and spring enlarged.

Bonnet Catch

To open the bonnet pull the control handle mounted centrally behind the fascia. The rear edge of the bonnet will rise an inch or so to be held by a spring loaded safety catch. Push back the catch and lift the bonnet, which can be held open by a rod, hinged to the bonnet surround and locating in the bonnet top.

When closing the bonnet, exert a slight downward pressure on the bonnet top until the locking catch is heard to engage.

DRIVING INSTRUCTIONS

Running In

It is most important to remember that at no time during the first 500 miles running-in period, must the engine be overloaded such as attempting to ascend steep inclines in top gear at low vehicle speeds. The load should be eased by changing down to a lower gear.

Fierce acceleration must also be avoided, and remember that the engine should never be raced in neutral.

On completion of the first 500 miles the running speed in each gear may be progressively increased, but full power should not be used until at least 1,500 miles have been covered, and even then only for short periods of time.

During this period a slight falling off in engine power may develop, in which case it will be beneficial to lightly grind in the valves and reset the valve clearance. No engine or complete car can be considered fully run-in until it achieves 2 to 3,000 miles.

The use of upper cylinder lubricant is advocated at all times, but most particularly during the first 2,000 miles. See Section 'K' for recommended brands.

Running-In Speed

2,500 r.p.m. for the first 1,000 miles.

Starting

Before starting the engine, see that the gear lever is in neutral and that the handbrake is applied.

Switch on the ignition and press the starter control firmly. Do not continue to press the starter button if the engine fails to start promptly. Allow a short interval between each successive attempt to start, and if the engine does not start in a reasonably short time, look for the cause of the trouble. Never press the starter button unless the engine is stationary.

As soon as the engine starts, release the starter button and push in the choke control to the half-way position. Release the choke completely as soon as the engine will run without it.

Do not allow the engine to race when first starting up as time must be allowed for the air to circulate properly. Let the engine idle fairly fast for a few minutes before moving off, and engage top gear as soon as

possible. Blanketing of the radiator will assist the engine to warm up quickly in cold weather, but always uncover the radiator before driving off. There is a thermostat to assist rapid warming up.

Driving

The gearbox has four forward speeds and reverse and incorporates a Laycock-de Normanville overdrive unit which gives two additional forward gear ratios. Start only in first gear, which is engaged by depressing the clutch pedal and moving the gear lever a little to the right and forwards as indicated on top of the gear lever (see fig. 2). Gradually release the clutch pedal, at the same time gently depressing the accelerator and releasing the handbrake. The car will move forward, gathering speed in accordance with the amount the accelerator is depressed.

Second gear is engaged by depressing the clutch pedal, moving the gear lever rearwards as far as it will go. Release the clutch pedal. Ease the accelerator when the higher gear is engaged.

To engage third gear, move the gear lever forwards, and then to the right, and forwards again as far as it will go.

High gear (top) is engaged by moving the gear lever as far as it will go rearward.

Changing down is affected by reversing the above procedure, with the exception that the accelerator pedal should be kept depressed whilst changing gear, in order to speed up the engine to suit the lower gear speed.

To engage reverse, which must only be done when the car is stationary, move the gear lever hard over to the right and rearwards. Remember, however, that the gearing is now lower than first gear. Therefore, release the clutch very slowly until the car just begins to move and then gently depress the accelerator to give the desired speed.

When temporarily halted on an incline do *not* slip the clutch—use the handbrake.

When descending a steep hill, it is advisable to engage a low gear as the engine will then provide a useful braking action.

Overdrive

Overdriving is entirely at the discretion of the driver and can be done in third and top gear by simply operating a small switch on the instrument panel.

On deceleration the overdrive remains in engagement. The overdrive will also remain in engagement even when the driver's switch is in the normal position unless the throttle is more than one-fifth open. This is due to the fact that a throttle switch is incorporated.

The fascia switch should not be moved into the normal position at speeds which are in excess of the direct drive maximum.

The following table gives the relationship between engine revolutions per minute to read speed in miles and kilometers per hour for the various gears. The top and second gear columns are divided to show the comparative engine revolutions with and without overdrive in operation.

Road Speed		Engine R.P.M					
K.P.H.	M.P.H.	1st	2nd	3rd	3rd + OD	Top	Top + OD
16	10	1710	1060	740	570	555	430
32	20	3415	2120	1480	1150	1110	860
48	30		3185	2220	1720	1665	1295
64	40		4245	2960	2295	2220	1725
80	50			3700	2870	2775	2155
96	60			4440	3440	3330	2585
112	70				4015	3885	3020
128	80				4590	4440	3450
144	90					4995	3880
160	100						4310
176	110						4740

Driving Hints

Do not press the starter control when a gear is engaged.

Remember to switch on the ignition before attempting starting the engine.

Refrain from continual pressing of the starter control if the engine will not fire.

Release the choke control as soon as possible after starting the engine.

Avoid leaving the car in gear with the handbrake off.

Never engage reverse gear when the vehicle is moving forward or a forward gear when the vehicle is moving backwards. Serious damage may result.

Do not slip the clutch in traffic or on an incline.

It is bad driving, apart from being injurious to the clutch, to coast the car with a gear engaged and the clutch pedal depressed.

Refrain from running the engine at high speeds for the first 500 miles.

Abstain from racing the engine in neutral at any time.

Remember not to run the car with the radiator completely blanked off.

Never fill the radiator with cold water when the engine is hot.

Do not under any circumstances run the engine in a closed garage or similar restricted atmosphere. The exhaust fumes are highly poisonous and if inhaled will quickly produce grave, if not fatal results.

GEARBOX

GENERAL DATA

Type	... Synchromesh on 2nd, 3rd, Top
Gear Control	Direct Gear Lever
Number of Gears	4 Forward, 1 Reverse
Type of Gears	Helical Constant Mesh
Oil Capacity (including overdrive)	5¼ Imp. pints; 6·3 U.S. pints; 2·98 litres

Gear Ratios :

1st Speed	3·073 : 1
2nd Speed	1·915 : 1
3rd Speed	1·332 : 1
4th Speed	Direct
Reverse	4·17 : 1

Overall Gear Ratios :

1st Speed	12·6
2nd Speed	7·85
3rd Speed	5·46
3rd Speed and Overdrive	4·24
Fourth Speed	4·1
Fourth Speed and Overdrive	3·18
Reverse	17·1

BEARINGS

Layshaft :

Type	Needle roller
Number of Rollers	46
Length of Roller	39·6 mm. (1·551 in.)
Diameter of Roller	3 mm. (·118 in.)

Mainshaft :

Make	R. & M.
Type	MJ. 35
Size	1·39 × 3·15 × ·827 in. (35 × 80 × 21 mm.)

First Motion Shaft :

Make	R. & M.
Type	1MJ4OG / / /
Size	1·58 × 3·55 × ·905 in. (40 × 90 × 23 mm.)

General Description

The gearbox has four forward speeds and one reverse, and synchromesh is incorporated on second, third and top gears.

Top gear is a direct drive; third and second are in constant mesh; first and reverse are obtained by sliding spur pinions.

Gear change is effected by means of a direct gear lever mounted on the gearbox side cover.

Lubrication

The gearbox oil level should be checked by the dipstick every 1,000 miles (1500 km.) and topped up if necessary.

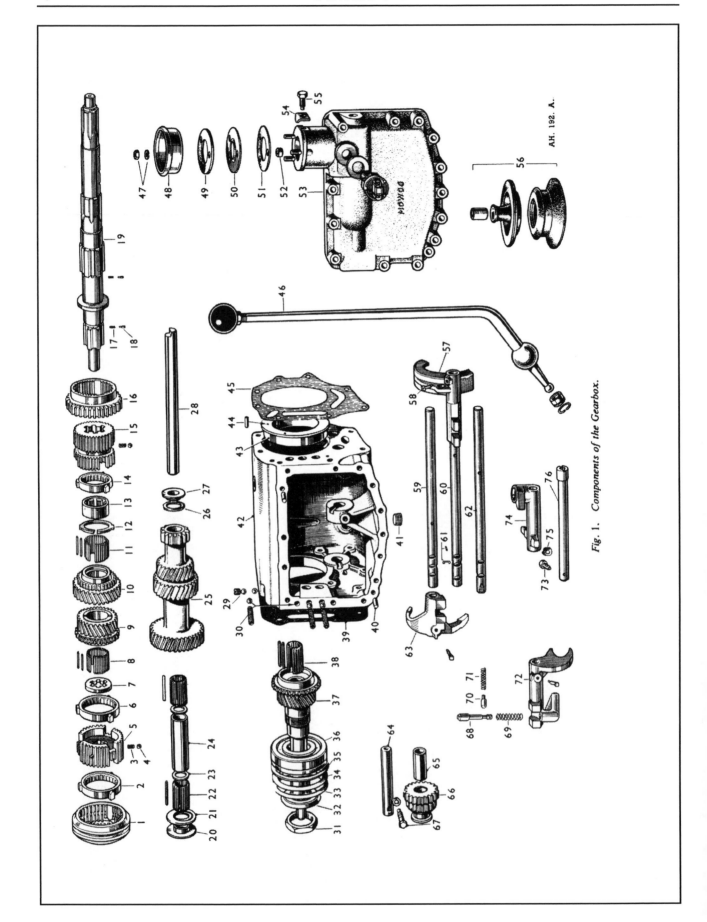

Fig. 1. Components of the Gearbox.

Caption for Fig. 1.

1. Synchromesh sleeve.	26. Washer.	52. Distance piece.
2. Baulking ring.	27. Thrust plate.	53. Side cover.
3. Synchronizer spring.	28. Layshaft.	54. Washer.
4. Synchronizer ball.	29. Interlocking balls.	55. Gear lever locating screw.
5. 3rd and 4th speed synchronizer.	30. Selector ball and spring.	56. Rubber dust covers.
6. Baulking ring.	31. Bearing nut.	57. 1st and 2nd speed fork.
7. Locking plate.	32. Bearing nut lockwasher.	58. Screw for fork.
8. Needle rollers.	33. Bearing spring plate.	59. 3rd and 4th speed fork rod.
9. Third speed gear.	34. Bearing plate.	60. 1st and 2nd speed fork rod.
10. Second speed gear.	35. Bearing circlip.	61. Interlocking pin and rivet.
11. Needle rollers.	36. First motion shaft bearing.	62. Reverse fork rod.
12. Gear washer.	37. First motion shaft.	63. 3rd and 4th speed fork.
13. Locking plate.	38. Needle rollers.	64. Reverse shaft.
14. Baulking ring.	39. Joint washer.	65. Bush.
15. 2nd speed synchronizer.	40. Side cover dowel.	66. Reverse gear.
16. First speed gear.	41. Drain plug.	67. Locking screw.
17. Plunger spring.	42. Gearbox casing.	68. Selector plunger.
18. Gear plunger.	43. Bearing housing.	69. Selector plunger spring.
19. Main shaft.	44. Locating peg.	70. Detent plunger.
20. Thrust plate.	45. Joint washer.	71. Detent plunger spring.
21. Thrust washer.	46. Gear lever.	72. Reverse fork.
22. Needle rollers.	47. Nut and washer.	73. Control shaft locating screw.
23. Washer, roller.	48. Cup.	74. Control lever.
24. Spacer, roller.	49. Rubber washer (thick).	75. Locking washer.
25. Laygear.	50. Steel washer.	76. Control shaft.
	51. Rubber washer (thin).	

The filler plug, which incorporates the dipstick, is located beneath a rubber cover, and is accessible when the floor mat and rubber cover have been raised.

After the first 500 miles the gearbox and overdrive should be drained and refilled with fresh oil. This procedure should be repeated afterwards every 6,000 miles (10000 km.).

Drain plugs are provided in the base of the gearbox and overdrive. Ensure that the hollow centre of the gearbox drain plug is kept clean. Do not forget to replace the plugs after draining.

The capacity of the gearbox plus overdrive is 5¼ pints (6·3 U.S. pints), (2·98 litres).

Removing the Gearbox

First disconnect the batteries by switching off the master switch at the rear left hand side of the boot compartment. Detach the starter lead at the starter.

Working within the car unscrew the gear lever knob and remove the locknut. For ease of working remove the seat cushions.

Remove the carpet covering the gearbox cover when access will be gained to the eight self tapping screws on each side flange of the cover. A further three screws will be found securing the forward end. The rubber grommet situated in the change speed lever aperture will require prising out and withdrawing up the lever. The gearbox cover may now be removed from the car.

Immediately in front of the cover there is a carpet covered bulkhead plate which can be removed for further access to the gearbox clutch housing. This plate is fixed to the bulkhead by six self-tapping screws.

Still within the car, disengage the propeller shaft from the driving flange by releasing the four securing nuts, bolts and washers. In order to turn the propeller shaft, to work on the nuts, the car may be pushed backwards or forward quite easily, providing the handbrake is off.

Next, disconnect the wires at their terminals on the overdrive switches. There is a main overdrive switch mounted on the gearbox side cover on which there are two wires. The overdrive solenoid switch is located on the left of the overdrive housing and only one wire has its terminal there. Release the union cap of the speedometer cable at the rear housing end, then extract the cable from the housing.

The two rear mounting points located on either side of the rear housing may now be released. Unscrew the two large nuts placed centrally to the brackets and not the four small setpins securing the brackets to the chassis.

Remove the various bolts and setpins securing the top of the clutch housing to the rear mounting plate. This includes the top bolt securing the starter to the gearbox.

As the following operations are performed under the gearbox it is necessary for the car to be placed over a servicing pit or the front of the car raised off the ground to a convenient working height.

Release the stabilising bar from the back of the gearbox by disconnecting it at the forward end.

G. 127. A

Fig. 2. Assembling gear and synchronizer.
1. *Gear.*   2. *Baulking ring.*   3. *Synchronizer.*

Disconnect the clutch linkage from the gearbox by first removing the split pin at the rear end of the adjustment bar and then unscrewing the bar away from the linkage. Detach the return spring placed immediately above it.

It is now necessary to lift the rear end of the engine in order to withdraw the gearbox. A hoist will be found suitable if a hook is secured to the rear end of the rocker cover. If a hoist is not available a suitable jack will give the required support. This method also relieves the load on the engine front mounting brackets when the gearbox and overdrive are removed.

Unscrew the remaining bolts and setpins holding the bottom of the clutch housing to the engine rear mounting plate, remembering that one of the bolts secures the starter. The latter can now be lifted out of the way.

The gearbox first motion shaft should now be withdrawn from the flywheel bearing and clutch by gently easing rearwards the gearbox and overdrive.

If the unit does not detach itself readily it is probably because the rear of the engine requires raising still further.

Replacing the gearbox and overdrive is a reversal of the removal procedure.

Dismantling the Gearbox

Remove the dipstick. Unscrew the breather from the overdrive unit. Drain the oil from the gearbox and overdrive by removing the drain plug beneath each unit.

Unscrew the speedometer drive from the right-hand side of the rear extension.

Unscrew the seven short and one long bolt and remove the clutch housing.

Remove the rubber cover located at the bottom of the gear lever when access will be gained to the three nuts threaded on studs mounted on the gear lever cup. With the removal of these nuts the cup may be withdrawn together with the three washers and three distance pieces located on the studs.

The gear lever can now be withdrawn from the gearbox.

Unscrew the thirteen bolts securing the side cover to the gearbox housing and remove the cover; there are two dowels locating the cover. Take care not to lose the three selector balls and springs which will be released as the cover is withdrawn.

Once the overdrive unit has been separated from the gearbox (see Service Manual), the removal of the adapter plate is accomplished by unscrewing the eight set pins mounted in the recess in the adapter plate.

The overdrive pump cam should slide freely along the third motion shaft thus giving access to the circlip holding the distance piece to the rear adapter plate. Remove the circlip and slide the distance piece off the shaft. The adapter plate should now pull away from the gearbox, together with the rear main bearing. It may be necessary for one operator to hold the gearbox vertically by the adapter plate whilst a second operator taps the third motion shaft until the ball race in the adapter plate is free of the shaft.

Cut the locking wires and unscrew the fork retaining screws. Remove the shifter shafts and forks in the following order :—

1. The reverse shaft and fork together with its selector and detent plungers and springs.

2. Top gear shifter shaft only.

3. First and second shaft and fork.

4. Top gear fork.

Take care not to lose the two interlock balls, normally located one at each side of the centre shifter shaft, which will be released when the shaft is removed.

Unscrew the reverse shaft locating screw and push out the shaft; lift the gear from the box.

Tap out the layshaft and allow the gear to rest in the bottom of the box.

Withdraw the drive gear assembly; note that there are 16 spigot rollers.

Withdraw the main shaft rearwards.

Lift out the layshaft gear and thrust washers.

Dismantling the Mainshaft

Slide the top and third gear hub and interceptors from the forward end.

Depress the plunger locating the third gear locking plate, rotate the plate to line up the splines and slide it from the shaft. Extract the plunger and spring, and slide off the third speed gear and its 32 rollers.

Depress the plunger locating the second gear locking collar and rotate the collar to line up the splines; slide off the collar, extract the plunger and spring and the two halves of the second speed washer; remove the gear with its 33 rollers from the shaft.

Slide the first and second speed hub, second speed interceptor and first speed gear rearwards from the shaft; if the first speed gear is withdrawn from the hub, take care to hold the balls and springs located in holes in the hub.

Depress the second gear locking collar plunger and rotate the collar to line up the splines; slide the collar from the shaft and extract the two halves of the second gear washer.

Withdraw the second speed gear and its 33 rollers from the shaft.

To dismantle the drive gear assembly, tap up the locking tab, unscrew the nut and remove the bearing.

Reassembly

Mainshaft: Smear the shaft with grease and assemble the 33 second speed gear rollers; slide the gear into position.

Replace the plunger and spring. Fit the two halves of the second gear washer and slide the collar on to the splines. Depress the plunger and push the collar into position, locating the lugs of the washer in the cut-outs of the collar; rotate the collar to bring the splines out of line.

Replace the balls and springs in the second and first speed hub; depress the balls and slide the first speed gear on to the hub; refit the assembly to the shaft.

Refit the bearing distance collar, the bearing and housing, the speedometer drive gear key and gear, locking washer and nut. Tighten the nut and tap over the locking washer.

Fit the third gear and its 32 rollers to the shaft; replace the plunger and spring and the third speed locking plate; rotate the plate to bring the splines out of line.

Fit the balls and springs to the top and third speed hub and slide the striking dog into position on the hub.

Replace the hub, striking dog and interceptors on the shaft.

Layshaft: Fit the distance tube to the layshaft gear with a washer at each end of the tube.

Smear the rollers with grease and position them in the gear. Place the thrust washers and plates in position at each end of the gear.

To retain the rollers in position, a length of round bar of layshaft diameter and just long enough to hold the thrust washers and plates, should be inserted in the gear assembly.

Place the gear in the box and allow it to rest at the bottom.

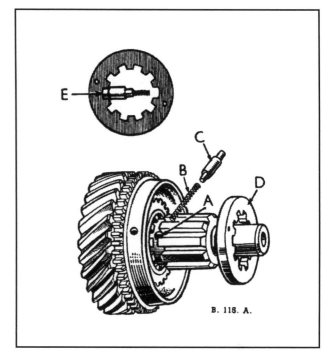

Fig. 3.   Securing the third motion shaft gears.
A. Hole for spring.   C. Location peg.
B. Spring.   D. Locking washer.
          E. Peg located in washer.

Reassembly of Gearbox

Insert the mainshaft assembly from the rear of the box.

Position the drive gear rollers and the drive gear assembly in the box.

Lift the layshaft gear into position, locating the thrust washer tags in the grooves provided. Push the layshaft through the housing and gear, and withdraw the retaining bar as the shaft pushes it out of the gear. The cut-away portion of the shaft must be aligned to fit the groove in the bell housing provided to prevent the layshaft from turning.

Refit the reverse gear and shaft and tighten the set-screw. Place the top gear shifter fork in the box. Replace the first and second gear shifter fork and shaft.

Replace one interlock ball above the first and second shifter shaft and insert the top gear shifter shaft.

Position the remaining interlock ball, holding it with grease and refit the reverse fork and shaft together with its selector and detent plungers and springs.

Screw in the fork set screws, tighten up and wire.

Slide the adapter plate, together with its bearing and paper joint washer, along the third motion shaft. Fit and tighten down the eight set pins securing the adapter plate to the gearbox.

Fit the distance piece which covers the space between the rear main bearing and the groove allocated for the circlip, and fix on the latter.

Refit the selector balls to the holes in the gearbox housing and the springs in the holes in the side cover.

The gear lever together with its cup, washers and distance pieces may now be attached to the side cover. Ensure that the ball of the lever makes a good fit with its mating socket.

Refit the cover, fitting a new gasket as required. Observe that the top right-hand setpin is longer than the other twelve.

Refit the clutch housing with plain bearing plate against the bearing.

Refit the speedometer drive, breather and dipstick.

OVERDRIVE

GENERAL DATA

Component	Dimensions New	Clearance New
Pump		
Plunger Diameter	$\frac{3}{8}$ in.—·0004 (9·525 mm.—·0102)	
	—·0008 (—·0203)	
Bore for Plunger	$\frac{3}{8}$ in.+·0008 (9·525 mm.+·0203)	+·0016 in. (·0406 mm.)
	—·0002 (—·0050)	+·0002 in. (·0051 mm.)
Plunger spring (fitted load at top of stroke)	9 lb. 12$\frac{3}{4}$ oz. (4·444 kg.)	
Valve spring load	5 lb. at $\frac{9}{16}$ in. long (2·268 kg. at 14·287 mm.)	
Pin for roller	$\frac{1}{4}$ in.±·00025 (6·35 mm.±·00635)	
Bore for pin in roller	$\frac{1}{4}$ in.+·002 (6·35 mm.+·0508)	+·00225 in. (·0572 mm.)
	+·001 (+·0254)	+·00075 in. (·0191 mm.)
Gearbox Third Motion Shaft		
Shaft diameter at steady bushes ...	1$\frac{5}{32}$ in.—·0009 (29·369 mm.—·0229)	
	—·0018 (—·0457)	
Steady bush internal diameter	1$\frac{5}{32}$ in.+·003 (29·369 mm.+·0762)	+·0048 in. (·1219 mm.)
	+·002 (+·0509)	+·0011 in. (·0279 mm.)
Shaft diameter at sun wheel bush ...	1$\frac{5}{32}$ in.—·0009 (29·369 mm.—·0229)	
	—·0018 (—·0457)	
Sun wheel bush internal	1$\frac{5}{32}$ in.+·002 (29·369 mm.+·0508)	+·0030 in. (·0762 mm.)
	+·003 (+·0762)	+·0009 in. (·0229 mm.)
Shaft diameter at rear steady bush ...	$\frac{5}{8}$ in.—·0008 (15·875 mm.—·0203)	
	—·0015 (—·0381)	
Rear steady bush internal diameter ...	$\frac{5}{8}$ in.+·001 (15·875 mm.+·0254)	+·0025 in. (·0635 mm.)
	—·000 (—·0000)	+·0008 in. (·0203 mm.)
Gear Train		
Planet pinion bearing	Torrington B78	
Planet bearing shaft external diameter...	$\frac{7}{16}$ in.+·0000 (11·11 mm. +·0000)	+·003 in. (·0762 mm.)
	—·0005 (—·0127)	+·0015 in. (·0381 mm.)
End float of sun wheel		·008 to ·014 in.
		(·2032 to ·3556 mm.)
Piston Bores		
Accumulator bores	1$\frac{1}{8}$ in.±·0005 (28·575 mm.±·0127)	
Operating piston bore	1$\frac{3}{8}$ in.±·0005 (34·125 mm.±·0127)	
Clutch		
Movement from direct to overdrive ...	·080 to ·120 in. (2·032 to 3·012 mm.)	
Allowance for wear direct drive clutch ...	— $\frac{1}{8}$ in. (3·175 mm.)	
Allowance for wear overdrive clutch ...	— $\frac{1}{8}$ in. (3·175 mm.)	

Lubrication

The lubricating oil in the overdrive unit is common with that in the gearbox and the level should be checked with the gearbox dipstick.

It is essential that an approved lubricant be used when refilling. Trouble may be experienced if some types of extreme pressure lubricants are used because the planet gears act as a centrifuge to separate the additives from the oil.

Recommended lubricants are given in section K. It should be emphasised that any hydraulically controlled transmission must have clean oil at all times and great care must be taken to avoid the entry of dirt whenever any part of the casing is opened.

Every 1,000 miles (1600 kilometres) check the oil level of the gearbox and overdrive and top up if necessary through the dipstick hole.

Every 6,000 miles (9600 kilometres) drain and refill the gearbox and overdrive unit. In addition to the normal drain plug fitted to the gearbox the overdrive unit incorporates a plug at its base which gives access to a filter. This plug should also be withdrawn to ensure that all used oil is drained away from the system.

Every 6,000 miles (9600 kilometers) after draining the oil, remove the overdrive oil pump filter and clean the filter gauze by washing in petrol. The filter is accessible through the drain plug hole and is secured by a central set bolt.

Refilling of the complete system (gearbox and overdrive) is accomplished through the gearbox filler plug. The capacity of the combined gearbox and overdrive unit is 5¼ pints (6·3 U.S. pints; 2·98 litres).

After draining, ¼ pint of oil will remain in the overdrive hydraulic system, so that only 5 pints will

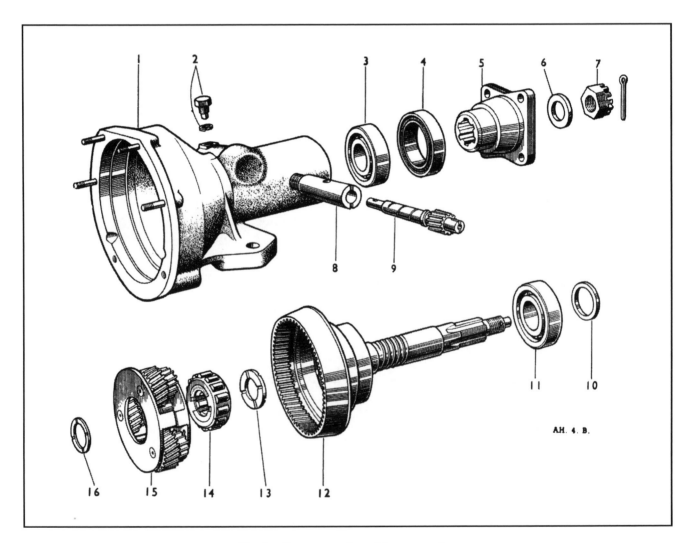

Fig. 1. Components of overdrive rear casing.

1. *Rear housing.*	5. *Driving flange.*	9. *Speedometer pinion.*	13. *Thrust washer.*
2. *Locking peg for '8'.*	6. *Washer.*	10. *Spacing washer.*	14. *Uni-directional clutch.*
3. *Outer bearing.*	7. *Flange nut.*	11. *Inner bearing.*	15. *Planet carrier.*
4. *Oil seal.*	8. *Speedometer pinion sleeve.*	12. *Annulus.*	16. *Spacing washer.*

be needed for refilling. If the overdrive has been dismantled the total of 5¼ pints will be required.

After refilling the gearbox and overdrive with oil, recheck the level after the car has been run, as a certain amount of oil will be retained in the hydraulic system of the overdrive unit.

PUMP VALVE

Access to the pump valve is gained through a cover on the left-hand side of the unit. Proceed as follows:—

Solenoid Operated Units
1. Remove drain plug and drain off oil.
2. Remove solenoid.
3. Slacken off clamping bolt in operating lever and remove lever, complete with solenoid plunger.
4. Remove distance collar from valve operating shaft.
5. The solenoid bracket is secured by two $\frac{5}{16}$ in. (7·938 mm.) studs and two $\frac{5}{16}$ in. diameter bolts, the heads of which are painted red, **remove the nuts from the studs before touching the bolts. This is important.** The two bolts should now be slackened off together, releasing the tension on the accumulator spring.
6. Remove the solenoid bracket.
7. Unscrew the valve cap and take out the spring, plunger and ball.

Reassembly is the reverse of the above operations. Ensure that the soft copper washer between the valve cap and pump housing is nipped up tightly to prevent oil leakage.

It will now be necessary to reset the valve operating lever. Proceed as follows:—

Before clamping up the valve operating shaft rotate the shaft until a $\frac{3}{16}$ in. (4·763 mm.) diameter pin can be inserted through the valve setting lever into the corresponding hole in the casing. Leave the pin in position, locking the unit in the overdrive position. Lift the solenoid plunger up to the full extent of its stroke (i.e. to its energised position) and clamp up the operating lever. The solenoid plunger bolt should now drop until it rests on the rubber stop immediately below. (Fig. 2). This stop gives the desired clearance between the plunger bolt and the boss situated on the solenoid bracket.

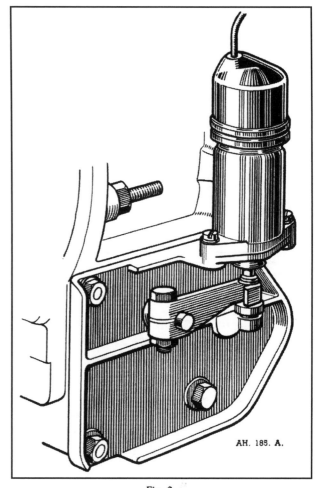

AH. 188. A.

Fig. 2.
Showing the solenoid plunger bolt resting in the required position on the rubber stop.

Remove the pin through the setting lever and operate the lever manually to check that the control operates easily.

HYDRAULIC PRESSURE

A working oil pressure of 420 to 440 lbs. per sq. in. (29·53 to 30·92 kg./cm².) is required.

The adaptor for use in connection with a pressure gauge is not now available from Messrs. Laycock Engineering Limited as stated in the Manual, but may be obtained from the Austin Motor Company Ltd. under the Service Tool Number 18G 251.

OVERDRIVE RELAY SYSTEM

Engagement of Overdrive is controlled electrically through a manually operated toggle switch. The circuit shown in Fig. 3 includes the following components :

(i) Relay, model SB40. An electro-magnetic switch used with item (ii) to enable an interlocking safeguard to be incorporated against changing out of overdrive with throttle closed.

(ii) Throttle Switch, model RTS1. A lever-operated semi-rotary normally closed switch used in conjunction with item (i) to override the toggle switch under closed throttle conditions.

(iii) Gear Switch, model SS10. A small plunger-operated switch allowing overdrive to be engaged only in the two highest forward-gear positions.

(iv) Solenoid Unit, model TGS1. An electro-magnetic actuator to engage overdrive mechanism by opening hydraulic control valve.

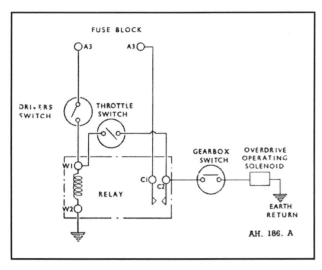

Fig. 3. Wiring diagram for the overdrive electrical circuit.

of the relay contacts connects terminal A3 to the gear switch and, providing one of the two higher ratio gears is engaged, will energize the solenoid unit and effect a change from direct drive to overdrive.

Overdrive will be maintained until the solenoid unit is de-energized.

Change from overdrive to direct drive is effected either by selecting a low gear (when the gear switch contacts will open) or by turning the toggle switch to off with open throttle (when the relay contacts will open).

If effected with closed throttle, a change from overdrive to direct drive could result in a shock to the transmission. An interlocking circuit is therefore incorporated to override the toggle switch under closed throttle conditions. Under these conditions, the throttle switch contacts provide an alternative supply circuit to the relay operating coil.

Operation

When the toggle switch contacts are closed, current flows by way of the ignition switch and fuse unit supply terminal A3 to energize the relay operating coil. Closure

Maintenance

Regular attention should be paid to wiring and connections. Damaged cabling must be replaced and loose terminals tightened, including the relay and solenoid unit earthing connections.

TRACING DEFECTS IN THE OVERDRIVE RELAY SYSTEM

The Solenoid Unit

With engine stopped and neutral gear engaged and ignition switched on, disconnect the solenoid connection. Using a jumper lead, momentarily connect the solenoid to fuse unit supply terminal A3. The solenoid should be heard to operate. If no sound is heard, the solenoid is defective or incorrectly adjusted to the operating linkage. Remake the connection.

The Gear Switch

Engage top gear, depress the throttle pedal and momentarily connect relay terminal C2 to terminal A3. The solenoid should be heard to operate. If no sound is heard, the gear switch is defective. Re-engage neutral gear.

The Relay Coil

Momentarily connect relay terminal W1 to terminal A3. The relay should be heard to operate. If no sound is heard, the relay is defective.

The Toggle Switch

Operate the toggle switch. The relay should be heard to operate. If no sound is heard, the toggle switch is defective.

The Relay Contacts

With top gear engaged, toggle switch closed and throttle switch open, the solenoid should be heard to operate. If no sound is heard, the relay is defective.

The Throttle Switch

Engage top gear and close toggle switch. Open toggle switch and slowly depress accelerator. The solenoid should be energized from zero to one-quarter throttle. If the solenoid is heard to release under one-quarter throttle, the throttle switch is defective.

FRONT HUBS AND INDEPENDENT FRONT SUSPENSION

GENERAL DATA

Hub Bearings :

Inner : Timken LM 67047 Taper Roller Bearing ... ... Size $\dfrac{1\cdot25}{1\cdot2506}\times\dfrac{2\cdot328}{2\cdot3286}\times\cdot660$ in.

Outer : Timken 07087X Taper Roller Bearing ... ... Size $\dfrac{\cdot875}{\cdot8756}\times\dfrac{1\cdot9687}{1\cdot9693}\times\cdot557$ in.

Caster Angle	$1\frac{3}{4}°$
Camber Angle	1°
Swivel Pin Inclination	$6\frac{1}{4}°$

Swivel Pin Diameter :

Top	$\cdot686\frac{1}{4}$ to $\cdot686\frac{3}{4}$ in. (17·43 to 17·44 mm.)
Bottom	$\cdot811\frac{1}{4}$ to $\cdot811\frac{3}{4}$ in. (20·6 to 20·61 mm.)

Swivel Pin Bush Tolerance

Top	$\cdot000\frac{3}{4}$ to $\cdot001\frac{3}{4}$ in. (·0019 to ·0045 mm.)
Bottom	$\cdot000\frac{3}{4}$ to $\cdot001\frac{3}{4}$ in. (·0019 to ·0045 mm.)

Independent Front Spring :

Free length	11·515 in. (29·25 cm.)
Fitted length	7·375 in. (18·72 cm.)
Number of Effective Coils	7
Diameter of Wire	·531 in. (13·48 mm.)
Inside Diameter of Coil	3·594 in. (9·11 cm.)

Shock Absorbers :

Make	Armstrong Hydraulic
Type	Double acting IS9/10RXP

FRONT HUBS

To Check for Wear

The inner and outer bearings of the front hub are of the taper roller type and are therefore adjustable. To check for wear of these bearings the car should be jacked up until the wheel of the front hub to be checked, is clear of the ground. Movement between the wheel and the back plate denotes wear of the hub bearings. Should a very positive movement be apparent, the front hub earings will need renewing.

Dismantling the Front Hubs

To dismantle either front hub, first jack up the car until the wheel is clear of the ground and then place blocks under the independent spring plate. Lower the car on to the blocks.

Remove the "knock-on" hub cap (direction of rotation marked on cap) and pull the wheel off the splines. Release the nuts and washers holding the brake drum, then gently tap the brake drum clear of the front

Fig. 1. Front Hub Exploded.

1. Grease cup.   6. Bearing outer race.
2. Axle nut.   7. Hub.
3. Split pin.   8. Bearing outer race.
4. Washer.   9. Inner bearing.
5. Outer bearing.   10. Oil seal.
            11. Swivel axle.

Insert shows distance piece and shims which are fitted to current models.

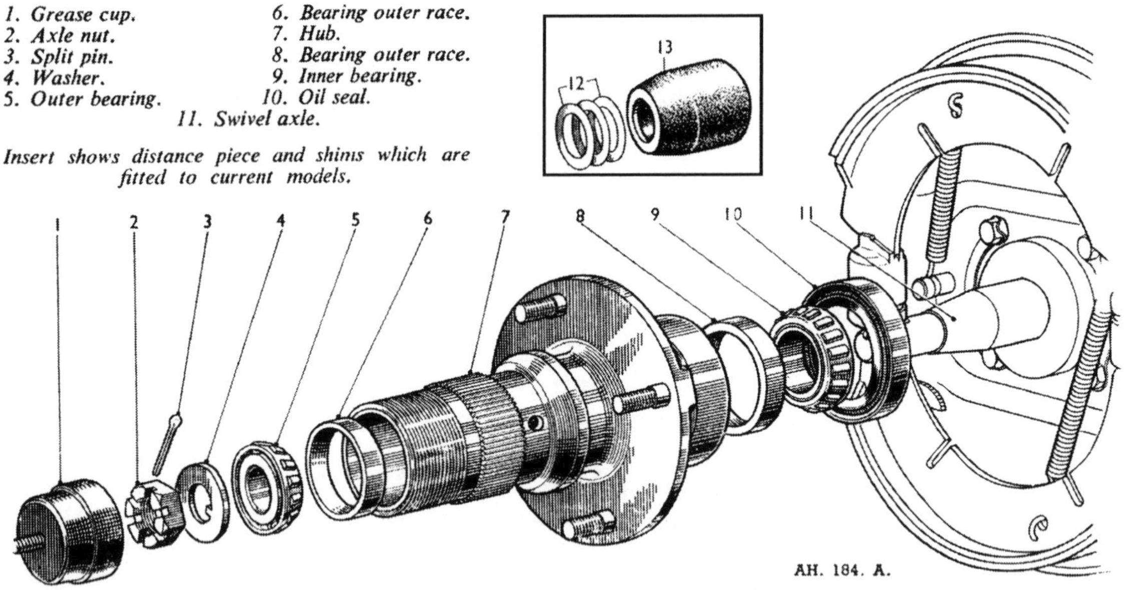

AH. 184. A.

hub assembly. If the drum appears to bind on the brake shoes, the shoe adjusters should be slackened.

Use the extractor provided in the tool kit to extract the grease retaining cup from within the hub. Straighten the end of the split-pin and then prize it out through the hole provided in the hub. Using a box spanner and tommy bar remove the hub securing nut and flat washer from the swivel axle.

The front hub can now be withdrawn by using an extractor. It is preferable to use an extractor which screws into position on the hub cap thread, but an extractor which locates over the hub studs may also be used. The hub is withdrawn complete with the inner and outer bearings and oil seal.

With the hub removed, the outer bearing can be dismantled by inserting a drift through the inner bearing and gently tapping the outer bearing clear of the hub. The inner bearing and oil seal can then be removed by inserting the drift from the opposite side of the hub.

Assembling the Front Hubs

When assembling the hub the inner taper roller bearing should first be inserted into the hub with the taper of the bearing facing inwards.

Pack the hub with *recommended* grease. Never use a grease thicker than that recommended.

Replace the outer bearing so that the taper of the bearing faces the inside of the hub. Use a soft metal drift to replace both bearings tapping them gently on diametrically opposite sides to ensure that they move

evenly into their respective housings in the hub. Replace the hub oil seal over the inner bearing so that the hollow side of the seal faces the bearing. Renew the seal if it is damaged in any way.

The hub can now be replaced on the swivel axle. This is best done by using a hollow drift which will bear evenly on both the inner and outer races of the outer hub bearing. Gently tap the hub into position until the inner race bears against the shoulder on the swivel axle.

Place the flat washer on the swivel axle and screw the nut down finger tight with the aid of a box spanner.

Replace the brake drum and secure with the four spring washers and self-locking nuts. Install the wheel.

Tighten the axle nut with a spanner, and at the same time rotate the wheel back and forth, until there is a noticeable drag. This ensures that the bearing cones are properly seated. The split pin should be inserted to lock the nut by unscrewing the axle nut to line up with the nearest split pin hole. The hub should now rotate freely, with no perceivable end play.

On later models a distance piece is fitted between each bearing. (See insert, Fig. 1). When adjusting make sure that the distance piece is a firm fit in the hub by packing with the necessary shims. These shims are selected to eliminate shock, therefore do not add too much thickness so as to cause undesired preload of the bearings.

Pack the retaining cup with grease and, using a drift, tap it gently but firmly up against the outer bearing. Replace the hub cap.

STEERING

GENERAL DATA

Make*	Cam Gears
Type of Gear	Cam and Lever
Steering Gear Ratio	12·6 to 1
Bearings	Ball Race and Felt Bush
Adjustment	Packing Shims
Diameter of Steering Wheel	17 ins. (43·18 cm.)
Turning Circle	35 ft. (10·668 m.)
Track Toe-in	...$\frac{1}{16}$ to $\frac{1}{8}$ in. (1·588 to 3·175 mm.)
Steering Connections	Ball and Socket Type

* Prior to Car Engine Number 231109 "Burman" Steering Gear was installed.

Description

The steering gear is a self-contained unit of extreme simplicity. The steering tube revolves a cam, which, in turn, engages with a taper peg fitted to a rocker shaft. This assembly is enclosed in an oil tight casing which carries two ball bearings at either end of the cam. These bearings are designed to carry radial and thrust loads.

When the steering wheel is turned the tube revolves the cam, which, in turn, causes the taper peg to move over a predetermined arc, thus giving the rocker shaft its desired motion. Attached to the rocker shaft is a steering side and cross tube lever, which links up with the steering linkage.

The steering is of the "three cross tube" type, having a centre cross tube connecting the steering side and cross tube lever to the arm on the idler shaft. Two shorter side tubes, one on either side, connect the steering arms to the steering gear and idler levers respectively.

Maintenance

Lubrication of the oil nipples on the steering connections and swivel bearings is most important to maintain accurate steering.

Approximately every 1,000 miles (1500 km.), use the oil gun with recommended oil to charge the following points with lubricant :—

(a) Steering rods and cross tube—6 nipples.
(b) Lower wishbone arm outer bearing—2 nipples.
(c) Swivel pin bushes—4 nipples.
(d) Steering idler—1 oil filler plug.

The steering box should be topped up with recommended oil to the top of the filler plug opening approximately every 1,000 miles (1500 km.).

Dismantling

Before the steering box can be dismantled, the steering arm must be removed. To do this, extract the split pin and unscrew the castellated nut, when the arm can be pulled off the splines by using a suitable extractor.

The top cover plate should be removed after extracting the four setscrews.

Turn the steering gear over and suitably support the top face leaving the rocker shaft free to be lightly tapped out using a soft metal drift. The follower peg is situated in the rocker and is a press fit. The peg is peined over at the top to ensure complete security. It should only be removed if showing an appreciable amount of wear.

Remove the four setpins securing the end cover plate in position, and release the end cover. The complete unit should now be up-ended with the steering box uppermost. By bumping the end of the inner shaft on a block of wood, placed on the floor, the worm with its two ball bearings will be displaced. The complete inner column can then be withdrawn from the casing through the open end of the steering box.

To extract the ball race at the top of the outer casing of the column pull upwards by hand, or if it is tight ease it from the column with a screwdriver behind the protruding lip. Replacing the ball race merely entails pushing it into place.

Assembling

Reassembly of the column is merely a reversal of the dismantling procedure, however, adjusting shims should be fitted behind the end cover so that there is no end play on the column, but at the same time they should not be preloaded, otherwise damage to the ball races may ensue.

When the rocker shaft is dropped into position, ensure that it is a good fit in its housing and that the oil seal at the lower end of the trunnion is making good contact.

Before refitting the top cover plate, screw back the adjuster. Ensure that all joints are oil tight.

Adjusting the Gear

The adjuster in the cover plate should be slackened by releasing the locknut and unscrewing the screw a few turns. Then the adjuster should be screwed down

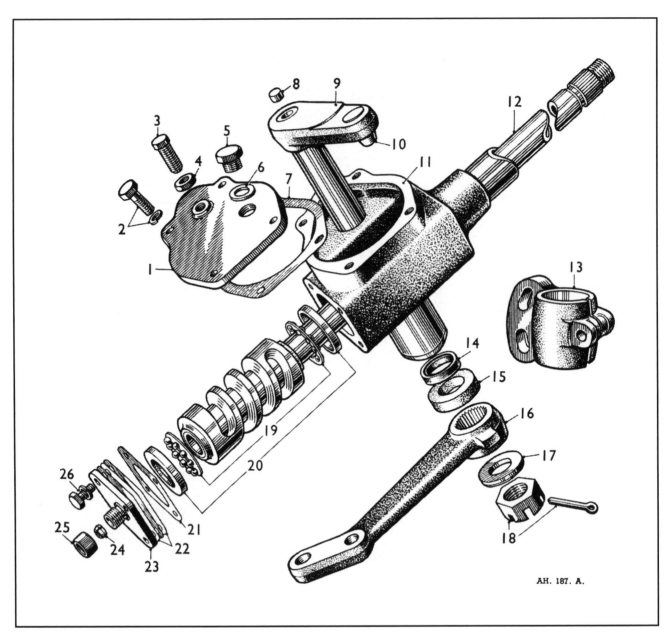

AH. 187. A.

Fig. 1. *Components of the steering box.*

1. *Top cover.*
2. *Setpin and washer.*
3. *Adjusting screw.*
4. *Locknut.*
5. *Filler plug.*
6. *Washer.*
7. *Joint washer.*
8. *Adjusting screw stop.*
9. *Follower peg screw.*
10. *Follower peg.*
11. *Steering box.*
12. *Inner column.*
13. *Steering box bracket.*
14. *Oil seal.*
15. *Dust excluder.*
16. *Steering lever.*
17. *Washer.*
18. *Castellated nut and washer.*
19. *Inner races.*
20. *Outer races.*
21. *Joint washer.*
22. *Adjusting shims.*
23. *End cover.*
24. *Olive.*
25. *Stator tube nut.*
26. *Setpin and washer.*

until there is no free movement in the straight ahead position of the gear and the adjustment secured by the locknut.

Final adjustment should be made once the gear has been reassembled to the body. It should be noted that as wear in use is normally greater in the straight ahead position than on lock, provision is made for this in the design of the cam, and it will be found that there is a slight end play towards each lock. It is essential, therefore, that adjustment should be made in the straight ahead position to avoid the possibility of tightness.

The steering gear should be filled with recommended gear oil via the filler plug, and then a final test made to ensure that the movement is free from lock to lock.

REAR AXLE

THE rear axle is of the three-quarter-floating type, incorporating hypoid final reduction gears. The axle shafts, pinion and differential assemblies can be withdrawn without removing the axle from the vehicle.

The rear axle wheel bearing outer races are located in the hubs; the inner races are mounted on the axle tube and secured by nuts and lock washers. Wheel studs in the hubs pass through the brake drums and axle shaft driving flanges. The brake drums are located on the hub flanges by two countersunk screws in each.

The differential and pinion shaft bearings are pre-loaded, the amount of pre-load being adjustable by shims. the position of the pinion in relation to the crown wheel is determined by a spacing washer.

Note:—The hypoid rear axle was introduced on the BN.1, commencing car engine number : 221536.

GENERAL DATA

Type	¾ floating
Ratio	4·1 : 1
Oil capacity	3 Imp. pints; 1·704 litres; 3·6 U.S. pints
Teeth in crown wheel	41
Teeth in Pinion	10
Final Drive	Hypoid
Crown wheel/pinion backlash	Marked on crown wheel

BEARINGS

Pinion Front

Make	Timken
Type	Taper Roller
Size	2·6875 × 1·125 × ·875 in. (68·26 × 28·58 × 22·2 mm.)

Pinion, Rear

Make	Timken
Type	Taper Roller
Size	3·125 × 1·375 × 1·1563 in. (79·38 × 34·92 × 29·37 mm.)

Differential

Make	R & M. LJT.45
Type	Double purpose Ball Journal
Size	85 × 45 × 19 mm.

Hub

Make	R & M. LDJ.45
Type	Ball Journal Double Row
Size	85 × 45 × 23 mm.

Lubrication

For the lubrication of the hypoid axle use lubricants only from approved sources, as tabulated in section "K". Do not, under any circumstances mix various brands of hypoid lubricant. If there is any doubt as to the oil previously used, drain and flush the axle with a little new hypoid oil before finally filling up. Do not use paraffin as a flushing medium. On a new car the oil should be drained and the axle refilled at 500 miles (750 km.) **and subsequently every 6,000 miles (10000 km.), failure to do this will eventually lead to axle break-down.**

The filler plug is situated on the rear side of the axle, and the drain plug in the bottom of the banjo casing.

AXLE UNIT REMOVAL

Loosen the hub caps, then jack-up the car and place supports under frame members just forward of rear springs front anchorage. Take off both wheels after removing the hub caps. Working from under the car, unscrew the four self-locking nuts and remove the bolts (U.N.F.) securing the propeller shaft flange to axle pinion flange.

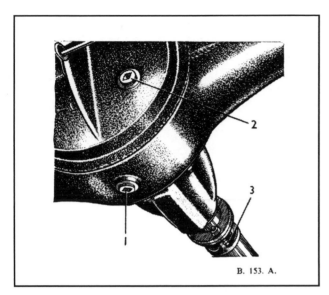

B. 153. A.

Fig. 1.   Rear Axle.
1. *Drain plug.*   2. *Filler plug.*
3. *Propeller shaft universal nipple.*

The handbrake cable should next be disconnected from the axle. This is accomplished by unscrewing it from its link to the brake balance lever, and unscrewing the nut holding its outer casing to the axle. The hydraulic brake pipe at the rear axle is detached from the flexible pipe at the union just forward of the right-hand shock absorber.

Unscrew the nuts securing the shock absorber links to the axle mounting brackets. Do not attempt to remove the links as this operation will prove much easier when freeing the axle.

The two rear axle bumpers, attached to the body in the form of metal boxes, must be detached by removing the four nuts and bolts found on each box. The heads of these bolts will be found within the boot.

Next remove the self-locking nuts from the spring clips ("U" bolts) which secure the axle to the springs. Observe that a fibre pad is situated between the axle and spring. Disconnection of the left-hand rear "U" bolt releases the anti-sway bar.

With the axle free the connecting links from the shock absorbers should be detached.

Remove the rubber block fixed between the axle and the left-hand chassis frame. It is not necessary to detach the corresponding block on the right-hand chassis frame.

The complete axle should be removed from the right-hand side of the car. Take care not to damage other components, particularly the petrol pump.

Installing the axle is the reverse of the above operations.

On re-assembling, it is advisable to jack-up the springs to meet the axle thus locating the spring centre bolt properly. Remember to fit the fibre pad.

When assembly is complete adjust the handbrake if required and bleed the hydraulic brake system all round.

MAINTENANCE

Whilst the following maintenance dismantling routines are described as for the axle being left in position on the car, the operator may remove the axle for his own convenience to work on such items as the gear carrier and bevel pinions on a bench.

Axle Shaft—To Remove and Replace

Loosen the hub cap of the wheel concerned before jacking-up the car. Remove the wheel after further unscrewing the hub cap, thus giving access to the rear hub extension. This is secured by five self-locking nuts.

Next, take out the two drum locating screws, using a screwdriver. The drum can be tapped off the hub and brake linings, provided the handbrake is released and the brake shoes are not adjusted so closely as to bind on the drum.

Should the brake linings hold the drum when the handbrake is released, it will be found necessary to slacken the brake shoe adjuster a few notches.

Remove the axle shaft retaining screw and draw out the axle shaft by gripping the flange outside the hub. It should slide easily but if it is tight on the studs it may need gently prising with a screwdriver inserted between the flange and the hub. Should the paper washer be damaged it must be renewed when re-assembling.

Replacement is a reversal of the above operations.

Make sure that the bearing spacer is in position.

Hub Removal and Replacement

Remove the drum and axle shaft. Remove the bearing spacer.

Knock back the tab of the locking washer and unscrew the nut with Service Tool 18G 258.

Tilt the lock washer to disengage the key from the slot in the threaded portion of the axle casing; remove the washer.

The hub can then be withdrawn with a suitable puller such as Service Tool No. 18G 220 with adaptors 'A', 'D' and 'E'. The bearing and oil seal will be withdrawn with the hub.

The bearing is not adjustable and is replaced in one straightforward operation.

When reassembling it is essential that the outer face of the bearing spacer should protrude from ·001 in. (·025 mm.) to ·004 in. (·091 mm.) beyond the outer face of the hub and the paper washer, when the bearing is pressed into position. This ensures that the bearing is gripped between the abutment shoulder in the hub and the driving flange of the axle shaft.

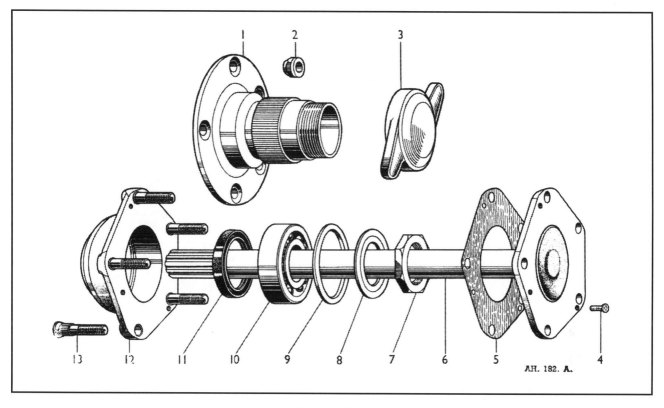

Fig. 2.   Rear Axle Hub Assembly.

1. *Hub extension.*	4. *Securing screw.*	7. *Hub locknut.*
2. *Securing nut.*	5. *Joint washer.*	8. *Hub lockwasher.*
3. *Hub cap.*	6. *Half shaft.*	9. *Bearing spacer.*

10. *Hub bearing.*	
11. *Oil seal.*	
12. *Hub casing.*	

13.   *Hub extension stud.*

Removing and Refitting Bevel Pinion Oil Seal

Mark the propeller shaft and the pinion driving flanges so that they may be replaced in the same relative position.   Disconnect the propeller shaft.

Unscrew the nut in the centre of the driving flange. Remove the nut and washer and withdraw the flange and pressed-on end cover from the pinion shaft.

Extract the oil seal from the casing.

Press a new oil seal into the casing with the edge of the sealing ring facing inwards.

Replace the driving flange end cover, taking care not to damage the edge of the oil seal.   Tighten the nut with a torque wrench to a reading of 1,680 lb. in. (19·36 kg. m.).

Reconnect the propeller shaft, taking care to fit the two flanges with the locating marks in alignment.

Removing the Differential

Drain the oil from the axle casing, and remove the axle shafts.

Mark the propeller shaft and pinion shaft driving flanges so that they may be replaced in the same relative positions; unscrew the self-locking nuts and separate the joint.

Unscrew the ten nuts securing the bevel pinion and gear carrier casing to the axle banjo; withdraw the casing complete with the pinion shaft and differential assembly.

Make sure that the differential bearing housing caps are marked so that they can be replaced in their original positions, then remove the four nuts and spring washers. Withdraw the bearing caps and differential assembly.

Remove the differential bearings from the differential case.   Note that the word "Thrust" is stamped on the thrust face of each bearing and that shims are fitted between the inner ring of each bearing and the differential case.

Knock back the tabs of the locking washers, unscrew the nuts from the bolts securing the crown wheel to the differential, and remove the crown wheel.

Tap out the dowel pin locating the differential pinion shaft.   The diameter of the pin is $\frac{3}{16}$-in. (4·8 mm.). The pinions and thrust washers can then be removed from the case.

Examination and Assembly

Examine the pinions and thrust washers and renew as required.

Examine the crown wheel teeth. If a new crown wheel is needed, a mated pair—pinion and crown wheel—must be fitted.

Replace the pinions, thrust washers and pinion shaft in the differential casing and insert the dowel pin. Peen over the entry holes.

Bolt the crown wheel to the differential case, but do not knock over the locating tabs. Tighten the nuts to a torque wrench reading of 540 lb. ft. (6·2 kg. m.).

Mount the assembly on two "V" blocks and check the amount of run out of the crown wheel as it is rotated, by means of a suitably mounted dial indicator.

The maximum permissible run out is ·002 in. (·05 mm.) and any greater irregularity must be corrected. Detach the crown wheel and examine the joint faces on the flange of the differential case and crown wheel for any particles of dirt.

When the parts are thoroughly cleaned it is unlikely that the crown wheel will not run true.

B. 134. A.

Fig. 4. Gauging the depth of the differential bearing housings.

B. 139. A.

Fig. 3. The Differential Carrier.

Tighten the bolts to the correct torque wrench reading and knock over the locking tabs.

Fit the differential bearings with the thrust faces outwards.

The Pinion Shaft

Remove the differential assembly. Unscrew the nut; remove the spring washer, the driving flange and the pressed end cover.

Drive the pinion shaft towards the rear; it will carry with it the inner race and the rollers of the rear bearing, leaving the outer race and the complete front bearing in position.

The inner race of the front bearing may be removed with the fingers after removal of the oil seal, and the outer race may be withdrawn with Service Tool No. 18G 264 with adaptors 'D' and 'H'.

Slide off the pinion sleeve and shims; withdraw the rear bearing inner race from the pinion shaft, noting the spacing washer against the pinion head.

Replacing Crown Wheel and Pinion

Fitting a new crown wheel and pinion involves four distinct operations :—
1. Setting the position of the pinion.
2. Adjusting the pinion bearing pre-load.
3. Adjusting the differential bearing pre-load.
4. Adjusting the backlash between the gears.

To carry out these operations correctly, three special tools are required; the bevel pinion setting gauge, Service Tool No. 18G 191, the pinion bearing outer race remover and replacer, Service Tool No. 18G 264 and the pre-load checking tool, Service Tool No. 18G 207.

Setting the Pinion Position
  (a) Fit the bearing outer races to the gear carrier.
  (b) Smooth off the pinion head with an oil stone, but do not erase the variation in pinion head thickness that is etched on the pinion head.
  (c) Refit the pinion head washer; if the original washer is damaged or not available, select a washer from the middle of the range of thicknesses: say, ·214 in. or ·216 in.
  (d) Fit the inner race of the rear bearing to the pinion shaft and position the pinion in the gear carrier without the shims, distance tube and oil seal. Fit the inner race of the front bearing.

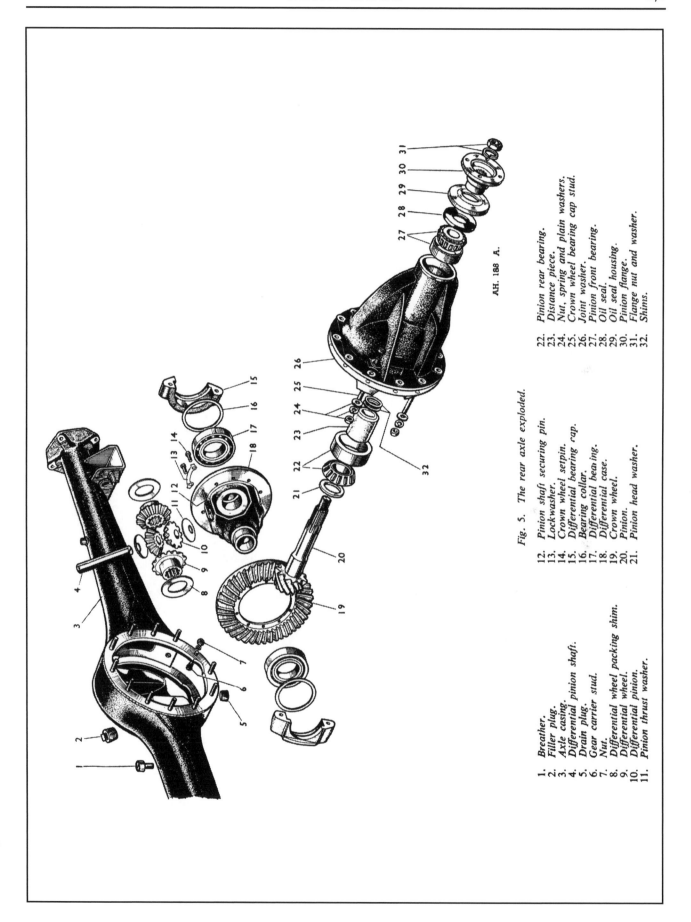

AH. 188 A.

Fig. 5. The rear axle exploded.

1. Breather.
2. Filler plug.
3. Axle casing.
4. Differential pinion shaft.
5. Drain plug.
6. Gear carrier stud.
7. Nut.
8. Differential wheel packing shim.
9. Differential wheel.
10. Differential pinion.
11. Pinion thrust washer.
12. Pinion shaft securing pin.
13. Lockwasher.
14. Crown wheel setpin.
15. Differential bearing cap.
16. Bearing collar.
17. Differential bearing.
18. Differential case.
19. Crown wheel.
20. Pinion.
21. Pinion head washer.
22. Pinion rear bearing.
23. Distance piece.
24. Nut, spring and plain washers.
25. Crown wheel bearing cap stud.
26. Joint washer.
27. Pinion front bearing.
28. Oil seal.
29. Oil seal housing.
30. Pinion flange.
31. Flange nut and washer.
32. Shims.

B. 138. A.

Fig. 6. Checking the bevel pinion bearing pre-load. Service Tool No. 18G 207.

(e) Refit the universal joint driving flange and tighten the nut gradually until a pre-load figure of 16 to 18 in. lb. (·184 to ·207 kg. m.) is obtained.

(f) Adjust the dial indicator to zero on the machined step "C" of the setting block.

(g) Remove the keep disc from the base of the magnet; clean the pinion head and place the magnet and dial indicator in position (Fig. 4). Move the indicator arm until the foot of the gauge rests on the centre of the differential bearing bore at one side and tighten the knurled locking screw. Obtain the maximum depth reading and note any variation from the zero setting.
Repeat the check in the opposite bearing bore. Add the two variations together and divide by two to obtain a mean reading.

(h) *With a standard pinion head (no variation marked).*
If the mean reading is within +·001 in. (·025 mm.) of the zero setting, the washer thickness is correct.
A positive mean reading indicates that the washer is not thick enough, and a negative mean reading indicates that it is too thick.
Example:

Thickness of washer fitted ...	·214 in.
Mean reading	+·003 in.
Thickness of washer required...	**·217 in.**

(i) *With a non-standard pinion head (variation marked).*
In addition to the procedure detailed above, allowance must also be made for the variation in thickness of the pinion head; a positive (+)

dimension must be subtracted from the thickness obtained above, and a negative (—) dimension added.
Using the same example and assuming a pinion head of non-standard thickness :
Example:

Thickness of washer fitted ...	·214 in.
Mean reading	+·003 in.
Total	**·217 in.**
Marked variation in pinion head thickness	+·002 in.
Thickness of washer required	**·215 in.**

A tolerance of ·001 in. is allowed in the thickness of the washer finally fitted.

Adjusting Pinion Bearing Pre-load

Assemble the pinion shaft bearings, distance tube, and shims to the gear carrier; fit the oil seal and driving flange. Tighten the flange nut gradually to a torque wrench reading of 1,680 lb. in. (19·4 kg. m.), checking the pre-load at intervals to ensure that it does not exceed 21 lb. in., i.e. 3 lb. in. greater than the previous figure as the oil seal is now fitted.

B. 137. A.

Fig. 7. Illustrating the machining tolerances for the differential bearing housings as marked by the factory inspector.

If the pre-load is too great more shims must be added, and if too small the thickness of the shimming must be decreased.

Adjusting the Differential Bearing Pre-load

Units marked with tolerances: The differential bearings must be pre-loaded and this is done by "pinching" them to the extent of ·002 in. on each bearing, the "pinch" being obtained by varying the thickness of the bearing distance collar fitted between each bearing outer ring and the register in the axle housing. The collar thickness is calculated as shown below.

In making the necessary calculations, machining tolerances and variations in bearing width must be taken into account. Machining tolerances are stamped on the component; bearing width variations must be measured.

The dimensions involved in pre-loading the differential bearings are illustrated in Fig. 8, and it is emphasised that it is the tolerance on each dimension which is important and referred to in the formula used.

The dimensions are:—

(A) From the centre line of the differential to the bearing register on the left-hand side of the gear carrier.
Tolerance: stamped on the carrier.

(B) From the centre line of the differential to the bearing register on the right-hand side of the carrier.
Tolerance: stamped on the carrier.

(C) From the bearing register on one side of the differential cage to the register on the opposite side.
Tolerance: stamped on the cage.

(D) From the rear face of the crown wheel to the bearing register on the opposite side.
Tolerance: stamped on the cage.

B. 138. A.

Fig. 9. Checking differential bearing width with Service Tool No. 18G 191.

To calculate the collar thickness :—

Left-hand side:

Formula: $A + D - C + \cdot 1815$ in. (4·610 mm.).

Substitute the dimensional tolerances for the letters in the formula. The result is the thickness of the collar required at the left-hand side to compensate for machining tolerances and to give the necessary pinch, *with bearings of standard width.* The width of the bearing must now be checked and any variation from standard added to or subtracted from the collar thickness. If the bearing width is under standard, that amount must be added to the collar thickness, and *vice versa.*

To check bearing width, rest the bearing on the small surface plate of Tool No. 18G 191 with the inner race over the recess and the thrust face downwards.

Place the magnet on the surface plate and set the dial indicator to zero on the step marked "C" of the small gauge block; this is the width of a standard bearing. Transfer the indicator to the plain surface of the bearing inner race and, holding the race down against the balls, note the reading on the dial. A **negative** reading shows the additional thickness to be **added** to the collar at this side; a **positive** reading, the thickness to be **subtracted.**

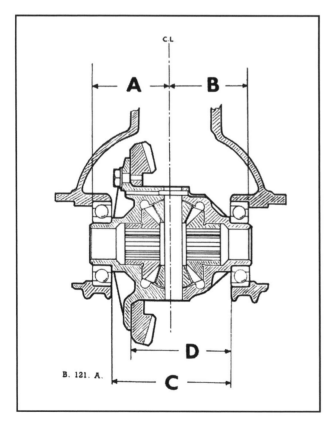

B. 121. A.

Fig. 8. Illustrates the points from which the calculations must be made to determine the shim thickness for the bearings on each side of the carrier.

Right-hand side:

    Formula: B—D+·1825 in. (4·634 mm.).

    The procedure is the same as that for the left-hand side.

Units not marked with tolerances: Some early models are fitted with differentials bearing no markings except the correct backlash for that particular pair of gears. The differential in such a case can be set as follows :—

(a) Fit the differential to the carrier with a distance collar at each side.

    By trial and error select collars of thicknesses such that the differential with bearings and collars just fits into the carrier without slack and without pinching the bearings.

(b) Remove the unit and add ·002 in. to the thickness of the collar at each side to give the required pre-load.

(c) Fit the unit to the carrier and bolt up.

(d) Check and adjust the backlash as detailed below

Table of Washer and Shim Thickness

Pinion head washer thicknesses	·208 in. to ·222 in. in steps of ·002 in.
Pinion bearing pre-load shims	·004 in. to ·012 in. in steps of ·002 in., plus ·020 in. and ·030 in.
Crown wheel bearing collars	·175 in. to ·185 in. in steps of ·002 in.
Pinion bearing pre-load ...	16 to 18 lb. in. without oil seal; 19 to 21 lb. in. with oil seal.
Crown wheel bearing pinch	·002 in. each side.

Adjusting Backlash

    Assemble the bearings to the differential cage and refit the differential to the gear carrier with the collars of calculated thickness.

    Mount the dial indicator on the magnet bracket so that an accurate measurement of the backlash can be taken. The recommended backlash is etched on the crown wheel.

B. 136. A.

Fig. 10.   *Checking crown wheel to pinion backlash   Service Tool Number* 18G 191.

    Vary the backlash by decreasing the thickness of the collar at one side and increasing the thickness of the collar at the other side by the same amount, thus moving the crown wheel into or out of mesh as required. The total thickness of the two collars must not be changed.

    A tolerance of +·002 in. (·05 mm.) to +·001 in. (·025 mm.) on the recommended backlash is allowable so long as this does not bring it below a minimum of ·006 in. (·152 mm.) or above a maximum of ·012 in. (·306 mm.).

BRAKES

A LTHOUGH the operation of the brakes on the BN.2 is similar to that of the BN.1, the larger and therefore redesigned front and rear brakes have made necessary the following additional information.

GENERAL DATA

Make	Girling
Type	Hydraulic, two leading shoes
Pedal free movement	$\frac{1}{8}$-in. (3·175 mm.)
Handbrake	Mechanical, rear wheels only
Drum diameter	11-in. (28 cm.)
Total frictional area	188 sq. in. (1213 sq. cm.)
Shoe lining width	$2\frac{1}{4}$-in. (57 mm.)
Shoe lining length :	
Front	10·4 in. (26·56 cm.)
Rear	10·4 in. (26·56 cm.)
Shoe lining thickness	·167 to ·174 in. (4·24 to 4·42 mm.)

Front Brakes

The front brakes are operated by two wheel cylinders situated diametrically opposite each other on the inside of the backplate and interconnected by a bridge pipe on the outside.

Each wheel cylinder consists of a light alloy body containing a spring seal support, seal, steel piston and edges of both shoes making initial contact with the drum. The shoes are allowed to slide and centralise during the actual braking operation which distributes the braking force equally over the lining area ensuring high efficiency and even lining wear.

Adjustment for lining wear is by means of two knurled snail cam adjusters, each operating against a peg at the actuating end of each shoe. Both adjusters turn clockwise to expand the shoes.

The brake shoes rest on supports formed in the backplate and are held in position by two return springs which pass from a hole in the abutment end of each web to a peg fixed to the backplate.

The bleed screw which is incorporated in one cylinder, is provided with a steel ball, this is normally seated firmly on a valve opening in the cylinder. A dust cover is fitted over the screw nipple to exclude dirt and with the removal of this cover and an anti-clockwise turn of the screw the fluid may escape.

Rear Brakes

The rear brake shoes are not fixed but are allowed to slide and centralise with the same effect as in the front brakes. They are hydraulically operated by a single acting wheel cylinder incorporating the handbrake

mechanism. At the cylinder end the leading shoe is located in a slot in the piston while the trailing shoe rests in a slot formed in the cylinder body. At the adjuster end they rest in slots in the adjuster links. Both shoes are supported on the backplate and are held in position by

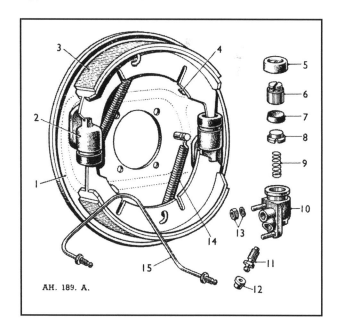

AH. 189. A.

Fig. 1. Front Brake Assembly.

1.	Backplate.	8.	Seal support.
2.	Wheel cylinder.	9.	Spring.
3.	Shoe.	10.	Cylinder housing.
4.	Snail cam adjuster.	11.	Bleed screw.
5.	Dust cover.	12.	Bleed screw cover.
6.	Piston.	13.	Nut and washer.
7.	Seal.	14.	Shoe return spring.
	15.	Cylinder connecting pipe.	

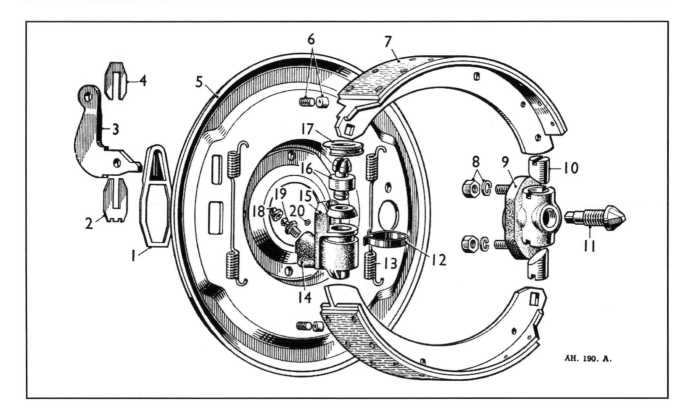

Fig. 2. Rear brake exploded.

1. Rubber seal.
2. Wheel cylinder locking plate.
3. Handbrake lever.
4. Wheel cylinder locking plate.
5. Backplate.
6. Steady post.
7. Brake shoe.
8. Nut and spring washer.
9. Adjuster body.
10. Adjuster tappets.
11. Adjuster wedge.
12. Dust cover clip.
13. Shoe return spring.
14. Pipe orifice.
15. Cylinder body.
16. Piston.
17. Dust cover.
18. Bleed nipple dust cover.
19. Bleed nipple.
20. Bleed valve ball.

two return springs fitted from shoe to shoe with the shorter spring nearer the adjuster.

The wheel cylinder consists of a light alloy die casting into the end of which moves a piston, with seal, in a highly finished bore. In to the other end of the housing is machined a slot to carry the trailing shoe and at right angles projecting through the backplate is pivoted the handbrake lever. The cylinder is attached to the backplate by a spring clip allowing it to slide laterally.

A bleed screw is incorporated in the cylinder housing with a rubber dust cap over the nipple end.

Adjustment for lining wear is made by the brake shoe adjuster. This has a steel housing which is spigotted and bolted firmly to the inside of the backplate. The housing carries two opposed steel links, the outer end

slotted to carry the shoes, and the inclined inner faces bearing on inclined faces of the hardened steel wedge.

The wedge has a threaded spindle with a square end which projects on the outside of the backplate, enabling a spanner to be used for adjustment purposes, by rotating the wedge in a clockwise direction, the links are forced apart and the fulcrum of the brake shoes expanded.

When the brake is applied, the piston under the influence of the hydraulic pressure, moves the leading shoe and the body reacts by sliding on the backplate to operate the trailing shoe.

The handbrake lever is pivoted in the cylinder body, and when operated the lever tip expands the trailing shoe, and the pivot moves the cylinder body and with it the piston to operate the leading shoe.

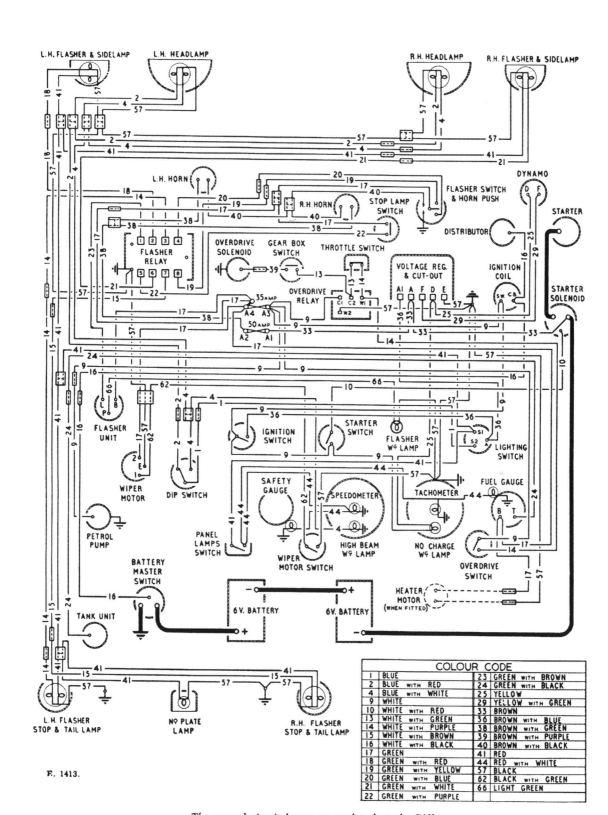

The general circuit layout as employed on the BN2.

F. 1413.

	COLOUR CODE			
1	BLUE	23	GREEN with BROWN
2	BLUE with RED	24	GREEN with BLACK
4	BLUE with WHITE	25	YELLOW
9	WHITE	29	YELLOW with GREEN
10	WHITE with RED	33	BROWN
13	WHITE with GREEN	36	BROWN with BLUE
14	WHITE with PURPLE	38	BROWN with GREEN
15	WHITE with BROWN	39	BROWN with PURPLE
16	WHITE with BLACK	40	BROWN with BLACK
17	GREEN	41	RED
18	GREEN with RED	44	RED with WHITE
19	GREEN with YELLOW	57	BLACK
20	GREEN with BLUE	62	BLACK with GREEN
21	GREEN with WHITE	66	LIGHT GREEN
22	GREEN with PURPLE		

		REGULAR ATTENTIONS	Section A Page No.
		DAILY.	
Oil.	A	Engine oil reservoir, check level	1
		1,000 MILES (1500 KM.)	
Oil.	B	Gearbox and Overdrive, top-up if necessary ...	1
		Rear Axle, top-up if required	1
		Steering Box, top-up	2
Oil Gun.	C	Propeller Shaft bearings and Splines (3) ...	2
		Swivel Axles (4). Lower suspension joints (2) ...	2
		Steering connections (6)	2
		Steering idler. 1 filler plug	2
		Shackle pins. Rear end of rear road springs (2)	2
		Clutch and Brake pedals pivot lever (1) ...	2
		Brakes balance lever (1), rear flexible cable (1)	2
Oil Can.	D	Steering column felt bush	2
		Handbrake, brake and clutch pedal linkages.	2
		Carburetter control joints	2
		Carburetters, replenish damper assembly oil reservoir	2
Examine.	E	Brake fluid supply tank, inspect and refill to correct level	2
		Shock absorbers, check for leaks	2
		3,000 MILES (5000 KM.)	
Oil.	F	Engine, drain and refill oil reservoir ...	2
Oil Can.	G	Distributor automatic advance, cam end drive shaft bearings, cam	3
		6,000 MILES (10000 KM.)	
Oil.	H	Gearbox and Overdrive, drain and refill ...	3
		Rear Axle, drain and refill	3
		Water Pump	4
		Air Cleaners, re-oil	4
Grease.	J	Front hub	3
		Dynamo bearings with H.M.P. grease ...	4
Examine.	K	Shock absorbers, top-up	4
		12,000 MILES (20000 KM.)	
Oil.	L	Engine Filter, renew element	5
Oil Gun.	M	Clutch operating shaft (2)	4
Grease.		Speedometer and Tachometer drives	4

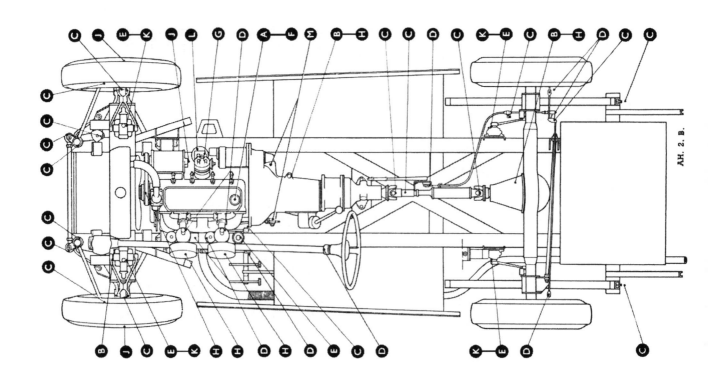

AH. 2. B.

RECOMMENDED LUBRICANTS

		Mobil	Shell	Wakefield	Esso	B.P.	Duckham's
Engine	Above 32° F. (0° C)	Mobiloil A	Shell X.100 30	Castrol XL	Esso Extra Motor Oil 20w/30	Energol Motor Oil S.A.E. 30	Duckham's "NOL THIRTY"
	32° F. Down to +10° F. (0° C. to -12° C.)	Mobiloil Arctic	Shell X.100 20/20W	Castrolite	Esso Extra Motor Oil 20w/30	Energol Motor Oil S.A.E. 20W	Duckham's "NOL TWENTY"
	Below 10° F. (-12° C.)	Mobiloil 10W	Shell X.100 10W	Castrol Z	Essolube 10	Energol Motor Oil S.A.E. 10W	Duckham's "NOL TEN"
Transmission		Mobiloil A	Shell X.100 30	Castrol XL	Essolube 30	Energol Motor Oil S.A.E. 30	Duckham's "NOL THIRTY"
Rear Axle and Steering Box		Mobilube GX. 90	Shell Spirax 90 E.P.	Castrol Hypoy	Esso ExPee Compound 90	Energol Transmission Oil E.P.—S.A.E. 90	Duckham's Hypoid 90
Oil Nipples		Mobilube GX. 140	Shell Spirax 140 E.P.	Castrol Hi-Press	Esso ExPee Compound 140	Energol Transmission Oil E.P.–S.A.E. 140	Duckham's NOL E.P. 140
Front Wheel Hubs		Mobilgrease M.P.	Shell Retinax A	Castrolease L.M.	Esso Multi-purpose Grease H	Energrease L.3	Duckham's L.B.10 Grease
Distributor and Oil Can ...		Mobil Handy Oil	Shell X.100 20/20W	Wakefield Everyman Oil	Esso Handy Oil	Energol Motor Oil S.A.E. 20W	Duckham's "NOL TWENTY"
Upper Cylinder Lubrication		Mobil Upperlube	Shell Donax U	Wakefield Castrollo	Esso Upper Cyl. Lubricant	Energol U.C.L.	Duckham's Adcoid Liquid

Rear Axle: For prevailing Sub-Zero 10° F. (-12° C.) temperatures use S.A.E. 80 Hypoid Lubricant.

Steering: For prevailing Sub-Zero 10° F. (-12° C.) temperatures use S.A.E. 80 Hypoid Lubricant.

Oil Nipples: For high temperature climates the grease as shown for hubs can be used.

Hydraulic Brakes: Use Girling Brake Fluid (Crimson).

Shock Absorbers: Use Armstrong's Super (Thin) Shock Absorber Oil.

Multigrade oils

In addition to the above oils we approve the use of multigrade oils produced by the companies shown, for all climatic temperatures, unless the engine is old or worn. Some are more expensive than the above oils because of their specific properties and greater fluidity at lower temperatures.

OFFICIAL TECHNICAL BOOKS

Brooklands Technical Books has been formed to supply owners, restorers and professional repairers with official factory literature.

Model	Original Part No.	ISBN
Workshop Manuals		
Austin-Healey 100 BN1 & BN2	97H997D	9780907073925
Austin-Healey 100/6 & 3000	AKD1179	9780948207471
(100/6 - BN4, BN6, 3000 MK. 1, 2, 3 - BN7, BT7, BJ7 & BJ8)		
Austin-Healey Sprite Mk. 1 Frogeye	AKD4884	9781855201262
Austin-Healey Sprite Mk. 2, Mk. 3 & Mk. 4 and	AKD4021	9781855201255
MG Midget Mk. 1, Mk. 2 & Mk. 3		
Parts Catalogues / Service Parts Lists		
Austin-Healey 100 BN1 & BN2	1050 Edition 3	9781783180363
Austin-Healey 100/6 BN4	AKD1423	9781783180493
Austin-Healey 100/6 BN6	AKD855 Ed.2	9781783180486
Austin-Healey 3000 Mk. 1 and Mk. 2 (BN7 & BT7)	AKD1151 Ed.5	9781783180370
Mk. 1 BN7 & BT7 Car no. 101 to 13750,		
Mk. 2 BN7 Car no. 13751 to 18888,		
Mk. 2 BT7 Car no. 13751 to 19853		
Austin-Healey 3000 Mk. 2 and Mk. 3 (BJ7 & BJ8)	AKD 3523 & AKD 3524	9781783180387
BJ7 Mk. 2 Car no. 17551 to 25314 and		
BJ8 Mk. 3 Car no. 25315 to 43026		
Austin-Healey Sprite Mk. 1 & Mk. 2 and	AKD 3566 & AKD 3567	9781783180509
MG Midget Mk. 1		
Austin-Healey Sprite Mk. 3 & Mk. 4 and	AKD 3513 & AKD 3514	9781783180554
MG Midget Mk. 2 & Mk. 3 (Mechanical & Body Edition 1969)		
Austin-Healey Sprite Mk. 3 & Mk. 4 and	AKM 0036	9780948207419
MG Midget Mk. 2 & Mk. 3 (Feb 1977 Edition)		
Handbooks		
Austin-Healey 100	97H996E	9781869826352
Austin-Healey 100/6	97H996H	9781870642903
Austin-Healey 3000 Mk 1 & 2	AKD3915A	9781869826369
Austin-Healey 3000 Mk 3	AKD4094B	9781869826376
Austin-Healey Sprite Mk 1 'Frogeye'	97H1583A	9780948207945

Also Available

Austin-Healey 100/6 & 3000 Mk. 1, 2 & 3 Owners Workshop Manual	9781783180455
Austin-Healey Sprite Mk. 1, 2, 3 & 4	9781855201255
MG Midget 1, 2, 3 & 1500 1958-1980 Owners Workshop Manual Glovebox Edition	
Austin-Healey Sprite Mk. 1, 2, 3 & 4	9781783180332
MG Midget 1, 2, 3 & 1500 1958-1980 Owners Workshop Manual	

Carburetters

SU Carburetters Tuning Tips & Techniques	9781855202559

Restoration Guide

Restoring Sprite & Midgets	9781855205987

Road Test Series

Austin-Healey 100 & 100/6 Gold Portfolio 1952-1959	9781855200487
Austin-Healey 3000 Road Test Portfolio	9791783180394
Austin-Healey Frogeye Sprite Road Test Portfolio 1958-1961	9781783180530
Austin-Healey Sprite Gold Portfolio 1958-1971	9781855203716

BROOKLANDS BOOKS LTD., P.O. BOX 904, AMERSHAM, BUCKS. HP6 9JA. UK
sales@brooklands-books.com
ISBN: 9780907073925   Part No. 97H997D   Ref: A09WH   4W4/

Printed in Great Britain
by Amazon

46569267R00130